Agronomy: Science and Technology

Agronomy: Science and Technology

Finn Cullen

SYRAWOOD
PUBLISHING HOUSE

New York

Published by Syrawood Publishing House,
750 Third Avenue, 9th Floor,
New York, NY 10017, USA
www.syrawoodpublishinghouse.com

Agronomy: Science and Technology
Finn Cullen

International Standard Book Number: 978-1-64740-058-3 (Hardback)

Cataloging-in-Publication Data

Agronomy : science and technology / Finn Cullen.
    p. cm.
Includes bibliographical references and index.
ISBN 978-1-64740-058-3
1. Agronomy. 2. Crops. 3. Soil management. 4. Agriculture. I. Cullen, Finn.
SB91 .A37 2022
631--dc23

Table of Contents

Preface

It is with great pleasure that I present this book. It has been carefully written after numerous discussions with my peers and other practitioners of the field. I would like to take this opportunity to thank my family and friends who have been extremely supporting at every step in my life.

The branch of science that focuses on producing and using plants for food, fiber, fuel and land reclamation is referred to as agronomy. It is a multi-disciplinary field that uses tools from various other fields such as plant genetics, meteorology and soil science. It also incorporates principles from different branches of science such as biology, chemistry, economics, earth science and genetics. Agronomy deals with a diverse range of issues like healthy food production, management of the environmental impact of agriculture as well as extracting energy from plants. The main areas that fall under agronomy are crop rotation, irrigation, soil classification, weed control, pest control and plant physiology. This textbook unfolds the innovative aspects of agronomy which will be crucial for the holistic understanding of the subject matter. Different approaches, evaluations and methodologies related to this field have been included in it. This textbook will serve as a reference to a broad spectrum of readers.

The chapters below are organized to facilitate a comprehensive understanding of the subject:

Chapter – What is Agronomy?

The science and technology of utilizing and producing plants for various agricultural purposes such as fuel, food, fiber and land restoration is referred to as agronomy. The chapter briefly introduces the key principles of agronomy as well as the related field of agricultural biotechnology to provide an extensive understanding of the subject.

Chapter – Agronomic Practices for Crop Production

The practices that are used to improve the soil quality and the environment, manage crops and enhance water usage are known as agronomic practices. All the diverse methods and practices used in agronomy, such as companion planting, shifting cultivation and crop rotation have been carefully analyzed in this chapter.

Chapter – Plant Breeding

The science of changing the characteristics of plants to increase their utility and value for humans is referred to as plant breeding. It is used to increase tolerance towards heat stress and drought stress in plants. This chapter discusses in detail the techniques related to breeding of plants to overcome these stresses as well as the selection methods which are used in this field.

Chapter – Agricultural Soil Science

The branch of soil science that focuses on the study of edaphic conditions with respect to the production of food and fiber is known as agricultural soil science. Some of the techniques studied within this field are contour plowing and mulching. The diverse applications of agricultural soil science and these techniques have been thoroughly discussed in this chapter.

Chapter – Agroecology

Agroecology is concerned with the study of ecological processes that are applied to agricultural production systems. Some of the diverse approaches towards agroecology are agro-population ecology, indigenous agroecology and inclusive agroecology. The topics elaborated in this chapter will help in gaining a better perspective about these approaches towards agroecology.

Chapter – Irrigation and its Types

Irrigation refers to the method of applying controlled amounts of water to plants at needed intervals to grow agricultural crops and maintain landscapes. The diverse methods of irrigation include micro irrigation, sprinkler irrigation and deficit irrigation. The topics elaborated in this chapter will help in gaining a better perspective about these methods of irrigation.

Finn Cullen

1

What is Agronomy?

The science and technology of utilizing and producing plants for various agricultural purposes such as fuel, food, fiber and land restoration is referred to as agronomy. The chapter briefly introduces the key principles of agronomy as well as the related field of agricultural biotechnology to provide an extensive understanding of the subject.

Agronomy embraces the branch of agriculture that deals with the development and practical management of plants and soils to produce food, feed, and fiber crops in a manner that preserves or improves the environment. The term "agronomy" represents the disciplines of soils, crops, and related sciences. In the soils area, specialties include soil microbiology, soil conservation, soil physics, soil fertility and plant nutrition, chemistry, biochemistry, and mineralogy. Specialties in the crops area relate primarily to plant genetics and breeding, crop physiology and management, crop ecology, turf-grass management, and seed production and physiology. Researchers in agronomy often work in close cooperation with scientists from disciplines such as entomology, pathology, chemistry, and engineering in order to improve productivity and reduce environmental problems. Even though less than 2 percent of the U.S. population are farmers who actively produce farm crops, the need for agronomists by other segments of society is increasing.

In the United States, field crops consist of those plants grown on an extensive scale, which differs from horticultural crops, which are usually grown intensively in orchards, gardens, and nurseries, but the distinctions are disappearing. Some of the major agronomic crops grown in the United States are alfalfa and pasture crops, peanuts, corn, soybeans, wheat, cotton, sorghum, oats, barley, and rice. Soil management aspects of agronomy encompass soil fertility, land use, environmental preservation, and non-production uses of soil resources for building, waste disposal, and recreation. Agronomists who work as soil scientists play extremely important roles in helping preserve water quality and preserve natural environments.

Agronomy is not a new field. As early as 7000 B.C.E. wheat and barley were grown at Jarmo, in present-day Iran. One could argue that the first farmers were in fact agronomists. In prehistoric times, humans shifted from foraging to cultivating specific crops, probably wheat or barley, for their food value. At harvest time, plants with easily gathered grain were selected first. This natural selection eventually made these food plants better adapted to continued cultivation because they were more easily harvested. Throughout the centuries, selection also occurred for other crop characteristics, such as taste, yield, and adaptation to specific soils and climates. The goal of today's production agronomists is essentially the same: to improve the quality, adaptability, and yield of our most important crops.

Science of Agronomy

There are both basic and applied aspects of agronomy. Agronomists examine very basic components of soils and crops at subcellular or molecular levels. For example, at the basic level, agronomists use sophisticated techniques to unravel the genetic makeup of major crops in order to change their adaptation, nutritive value, or to breed medicinal benefits into agronomic crops. Genetic improvement is an area where major breakthroughs are likely to occur. Agronomists have developed highly specialized computer models of crop growth in order to better understand how environmental and management components affect the way crops grow. These models help in the development of such things as precision fertilizer application techniques, which provide the crop with the correct amount of nutrients at the correct time in its life cycle. This technique helps reduce fertilizer overapplication, which is costly to the farmer, and may increase groundwater pollution. Models of how chemicals move in the soil also help assure proper application of animal manures, municipal waste, and soil amendments necessary for crop growth. Molecular components of soil constituents are studied to determine basic interactions affecting plant growth and nutrition, and soil and water quality.

International Agronomy

Agronomy is an international discipline. Many of the problems, issues, and challenges faced by societies around the world are universal in nature, and require international cooperation. For example, a major problem facing the developed world is that of how best to use our land resources. Within the developing world, the same problems exist. The questions of how much and which land should be saved for food and fiber production and which land should be used for nonagricultural uses must be addressed by both developing and developed societies. Agronomists play a crucial role in assessing land quality to assure an environmentally friendly use of land. Studying how plants adapt to differing climates and environments has allowed plant scientists to increase food and fiber production in regions of the world where the necessities of life are most limited. Knowledge gained and disseminated by agronomists in the developed world has helped improve the human condition in the developing world. For example, plant geneticists and breeders use similar hybrid and variety development techniques in both developed and developing countries. Through plant breeding, for example, agronomists have developed high-yielding rice that is adapted to tropical climates. Breakthroughs in gene transfer permit plant breeders to improve grain quality and nutritional traits. These techniques have also contributed to increased production efficiency by genetically incorporating into food crops increased pest resistance and by broadening their range of adaptation.

Principles of Agronomy

Agronomic principles are the ways and means for the better management of soil, plants and environment for economically maximum returns per unit area. Principles of crop management depends largely on the type of farming namely, specialized, diversified, mixed and integrated and also on the physical and technological facilities available, irrigated farming, dry farming and rain fed farming.

The fundamental principles of agronomy may be listed as:

1. Planning, programming and executing measures for maximum utilization of land, labor, capital, sunshine, rain-water, temperature, humidity, transport and marketing facilities.

2. Choice of crop verities adaptable to the particular agro-climate, land situation, soil fertility, season and method of cultivation and befitting (suitable) to the cropping systems.

3. Proper field management by tillage, preparing field channels and bunds for irrigation and drainage, checking soil erosion, leveling and adopting other suitable land improvement practices.

4. Adoption of multiple cropping and also mixed or inter cropping to ensure harvest even under adverse environmental conditions.

5. Timely application of proper and balanced nutrients to the crop or crops in sequence and improvement of soil fertility and productivity. Correction of bad effect of soil reactions and conditions and increasing soil organic matter through the application of green manner. FYM, organic wastes, bio-fertilizer and profitable recycling of organic wastes.

6. Choice of quality seed or seed material and maintenance of requisite plant density per unit area with healthy and uniform seedlings.

7. Proper water management with respect to crop, soil and environment through conservation and utilization of soil moisture as well as by water that is available in excess.

8. Adoption of adequate, need-based, timely and proper plant protection measures against weed's insect-pests, pathogens, as well as climatic hazards and correction of deficiencies and disorders.

9. Adoption of suitable and proper management practices including intercultural operations to get maximum benefit.

10. Adoption of suitable method and time of harvesting of crop to reduce field-damage and to release land for succeeding crop(s) and efficient utilization of residual moisture, plant nutrients and other management practices.

11. Adoption of suitable post harvest technologies.

PLANT POPULATION

Plant population is defined as the total number of plants present at unit area of land, while plant spacing is the arrangement of plants on an area. The yield of crop is directly influenced by population of plant. Actually the yield of a crop is the end result of final plant population which is influenced by the number of viable seed germination and survival rate. The population of plant should not so much high that can drain out most of the moisture from the field before the crop reaches to maturity stage. As well as population should not too low that moisture remain unutilized.

If the concentration of plants is high then certain alterations in growth of plant can be observe. As plant height is increased, in high plant concentration due to the competition for light. The thickness of leave may reduce due to high plant population and leaf geometry is also altered due to high population pressure. High plant density cause low yield of individual plant due to reduction in number of ears in indeterminate plants. While reduced yield in determinate plants is due to small size of ear panicles.

Crop stand establishment plays a key role in crop life cycle and can lead to a well-established planting, dynamic growth and ultimately high production of the crop. The aim of establishing a good crop stand is to attain Maximum and uniform germination and emergence vigorous seedlings, these vigorous seedlings may also perform well under stress conditions. Planting time, land preparation, sowing methods and seed quality are some important factors which can affect the population of plant.

High and better crop establishment rates are keys for optimum plant population, uniform growth and maturity, crop's resistance to diseases and insects, competing with weeds and optimizing yield. Good crop stand establishment is essential for the efficient use of water, light and available resources.

The quality of seed is the most important factor which affects plant population. Seed should always be disease free to ensure optimum plant population. High plant population enhances the interplant competition for nutrients water and light which affect the yield due to the stimulation of apical dominance and induces barrenness which ultimately decreases ear production per plant in maize.

The population of plants should always keep optimum. Optimum plant population insures maximum crop yield. Soil cover and the size of plant are the factors that influence optimum plant population. Following are the approaches that help to maintain optimum population of plants:

- Preparation of seed bed.
- Arrangement of rows and beds.
- Maintenance of row spacing.
- Accurate planting time and methods.
- Seed of good quality and health.
- Insuring maximum germination percentage.

Crop productivity is usually affected by limiting resources and it determined how the crop utilizes these resources. Crop productivity can be enhanced by maintaining optimum plant population in the field.

The global requirement of agricultural crops for food is increasing at a rapid pace. Global food production must be increased between 50 and 70% by 2050 to meet the projected population growth. The possible options to fulfill this food need are to increase area of production or to increase crop production per unit area. Increasing productivity on existing agricultural land is preferable as it avoids greenhouse gas emissions and the large-scale disruption of existing ecosystems associated with bringing new land into production.

AGRICULTURAL BIOTECHNOLOGY

Agricultural biotechnology is a range of tools, including traditional breeding techniques, that alter living organisms, or parts of organisms, to make or modify products; improve plants or animals; or develop microorganisms for specific agricultural uses. Modern biotechnology today includes the tools of genetic engineering.

Uses of Agricultural Biotechnology

Biotechnology provides farmers with tools that can make production cheaper and more manageable. For example, some biotechnology crops can be engineered to tolerate specific herbicides, which make weed control simpler and more efficient. Other crops have been engineered to be resistant to specific plant diseases and insect pests, which can make pest control more reliable and effective, and/or can decrease the use of synthetic pesticides. These crop production options can help countries keep pace with demands for food while reducing production costs. A number of biotechnology-derived crops that have been deregulated by the USDA and reviewed for food safety by the Food and Drug Administration (FDA) and/or the Environmental Protection Agency (EPA) have been adopted by growers.

Many other types of crops are now in the research and development stages. While it is not possible to know exactly which will come to fruition, certainly biotechnology will have highly varied uses for agriculture in the future. Advances in biotechnology may provide consumers with foods that are nutritionally-enriched or longer-lasting, or that contain lower levels of certain naturally occurring toxicants present in some food plants. Developers are using biotechnology to try to reduce saturated fats in cooking oils, reduce allergens in foods, and increase disease-fighting nutrients in foods. They are also researching ways to use genetically engineered crops in the production of new medicines, which may lead to a new plant-made pharmaceutical industry that could reduce the costs of production using a sustainable resource.

Genetically engineered plants are also being developed for a purpose known as phytoremediation in which the plants detoxify pollutants in the soil or absorb and accumulate polluting substances out of the soil so that the plants may be harvested and disposed of safely. In either case the result is improved soil quality at a polluted site. Biotechnology may also be used to conserve natural resources, enable animals to more effectively use nutrients present in feed, decrease nutrient runoff into rivers and bays, and help meet the increasing world food and land demands. Researchers are at work to produce hardier crops that will flourish in even the harshest environments and that will require less fuel, labor, fertilizer, and water, helping to decrease the pressures on land and wildlife habitats.

In addition to genetically engineered crops, biotechnology has helped make other improvements in agriculture not involving plants. Examples of such advances include making antibiotic production more efficient through microbial fermentation and producing new animal vaccines through genetic engineering for diseases such as foot and mouth disease and rabies.

Benefits of Agricultural Biotechnology

The application of biotechnology in agriculture has resulted in benefits to farmers, producers, and consumers. Biotechnology has helped to make both insect pest control and weed management safer and easier while safeguarding crops against disease.

For example, genetically engineered insect-resistant cotton has allowed for a significant reduction in the use of persistent, synthetic pesticides that may contaminate groundwater and the environment.

In terms of improved weed control, herbicide-tolerant soybeans, cotton, and corn enable the use of reduced-risk herbicides that break down more quickly in soil and are non-toxic to wildlife and humans. Herbicide-tolerant crops are particularly compatible with no-till or reduced tillage agriculture systems that help preserve topsoil from erosion.

Agricultural biotechnology has been used to protect crops from devastating diseases. The papaya ringspot virus threatened to derail the Hawaiian papaya industry until papayas resistant to the disease were developed through genetic engineering. This saved the U.S. papaya industry. Research on potatoes, squash, tomatoes, and other crops continues in a similar manner to provide resistance to viral diseases that otherwise are very difficult to control.

Biotech crops can make farming more profitable by increasing crop quality and may in some cases increase yields. The use of some of these crops can simplify work and improve safety for farmers. This allows farmers to spend less of their time managing their crops and more time on other profitable activities.

Biotech crops may provide enhanced quality traits such as increased levels of beta-carotene in rice to aid in reducing vitamin A deficiencies and improved oil compositions in canola, soybean, and corn. Crops with the ability to grow in salty soils or better withstand drought conditions are also in the works and the first such products are just entering the marketplace. Such innovations may be increasingly important in adapting to or in some cases helping to mitigate the effects of climate change.

The tools of agricultural biotechnology have been invaluable for researchers in helping to understand the basic biology of living organisms. For example, scientists have identified the complete genetic structure of several strains of Listeria and Campylobacter, the bacteria often responsible for major outbreaks of food-borne illness in people. This genetic information is providing a wealth of opportunities that help researchers improve the safety of our food supply. The tools of biotechnology have "unlocked doors" and are also helping in the development of improved animal and plant varieties, both those produced by conventional means as well as those produced through genetic engineering.

Safety Considerations with Agricultural Biotechnology

Breeders have been evaluating new products developed through agricultural biotechnology for centuries. In addition to these efforts, the United States Department of Agriculture (USDA), the Environmental Protection Agency (EPA), and the Food and Drug Administration (FDA) work to ensure that crops produced through genetic engineering for commercial use are properly tested and studied to make sure they pose no significant risk to consumers or the environment.

Crops produced through genetic engineering are the only ones formally reviewed to assess the potential for transfer of novel traits to wild relatives. When new traits are genetically engineered into a crop, the new plants are evaluated to ensure that they do not have characteristics of weeds. Where biotech crops are grown in proximity to related plants, the potential for the two plants to exchange traits via pollen must be evaluated before release. Crop plants of all kinds can exchange traits with their close wild relatives (which may be weeds or wildflowers) when they are in proximity. In the case of biotech-derived crops, the EPA and USDA perform risk assessments to evaluate this possibility and minimize potential harmful consequences, if any.

Other potential risks considered in the assessment of genetically engineered organisms include any environmental effects on birds, mammals, insects, worms, and other organisms, especially in the case of insect or disease resistance traits. This is why the USDA's Animal and Plant Health

Inspection Service (APHIS) and the EPA review any environmental impacts of such pest-resistant biotechnology derived crops prior to approval of field-testing and commercial release. Testing on many types of organisms such as honeybees, other beneficial insects, earthworms, and fish is performed to ensure that there are no unintended consequences associated with these crops.

With respect to food safety, when new traits introduced to biotech-derived plants are examined by the EPA and the FDA, the proteins produced by these traits are studied for their potential toxicity and potential to cause an allergic response. Tests designed to examine the heat and digestive stability of these proteins, as well as their similarity to known allergenic proteins, are completed prior to entry into the food or feed supply. To put these considerations in perspective, it is useful to note that while the particular biotech traits being used are often new to crops in that they often do not come from plants (many are from bacteria and viruses), the same basic types of traits often can be found naturally in most plants. These basic traits, like insect and disease resistance, have allowed plants to survive and evolve over time.

Uses of Biotechnology Crops

According to the USDA's National Agricultural Statistics Service (NASS), biotechnology plantings as a percentage of total crop plantings in the United States in 2012 were about 88 percent for corn, 94 percent for cotton, and 93 percent for soybeans. NASS conducts an agricultural survey in all states in June of each year. The report issued from the survey contains a section specific to the major biotechnology derived field crops and provides additional detail on biotechnology plantings.

The USDA does not maintain data on international usage of genetically engineered crops. The independent International Service for the Acquisition of Agri-biotech Applications (ISAAA), a not-for-profit organization, estimates that the global area of biotech crops for 2012 was 170.3 million hectares, grown by 17.3 million farmers in 28 countries, with an average annual growth in area cultivated of around 6 percent. More than 90 percent of farmers growing biotech crops are resource-poor farmers in developing countries. ISAAA reports various statistics on the global adoption and plantings of biotechnology derived crops.

REGENERATIVE AGRICULTURE

Regenerative agriculture is a conservation and rehabilitation approach to food and farming systems. It focuses on topsoil regeneration, increasing biodiversity, improving the water cycle, enhancing ecosystem services, supporting biosequestration, increasing resilience to climate change, and strengthening the health and vitality of farm soil. Practices include, recycling as much farm waste as possible, and adding composted material from sources outside the farm.

Regenerative agriculture on small farms and gardens is often based on ideologies like permaculture, agroecology, agroforestry, restoration ecology, keyline design and holistic management. Large farms tend to be less ideology driven, and often use "no-till" and/or "reduced till" practices.

On a regenerative farm, yield should increase over time. As the topsoil deepens, production may increase and less external compost inputs are required. Actual output is dependent on the nutritional value of the composting materials, and the structure and content of the soil.

Roots

Rodale Institute, Test Garden.

Regenerative agriculture is based on various agricultural and ecological practices, with a particular emphasis on minimal soil disturbance and the practice of composting. Maynard Murray had similar ideas, using sea minerals. Her work led to innovations in no-till practices, such as slash and mulch in tropical regions.

Field Hamois Belgium Luc Viatour.

The Lasagna method, also called sheet mulching, is a regenerative agriculture practice that feeds the soil biome from above and encourages the soil food web to aerate and mix the nutrients from the mulch, into the soil below.

Microbiologist, Elaine Ingham, popularized among farmers and gardeners, the importance of soil health and the soil food web in her book, Soil Biology Primer, in 2000.

"By marching forward under the banner of sustainability we are, in effect, continuing to hamper ourselves by not accepting a challenging enough goal. I am not against the word sustainable, rather I favor regenerative agriculture."

— Robert Rodale.

Agroforestry on a grazing farm.

However, the institute stopped using the term in the late 1980's, and it only appeared sporadically (in 2005 and 2008), until they released a white paper in 2014, titled "Regenerative Organic Agriculture and Climate Change". The paper's summary states, "we could sequester more than 100% of current annual CO_2 emissions with a switch to common and inexpensive organic management practices, which we term 'regenerative organic agriculture.'" The paper described agricultural practices, like crop rotation, compost application, and reduced tillage, that are similar to organic agriculture methods.

From 1990 to 2010, regenerative agriculture was primarily practiced within the permaculture community. The ecological systems approach to permaculture, influenced by Carol Sanford and the design and development work of the Regenesis group, led regenerative agriculture to incorporate whole farm design, multi-story agroforestry and livestock rotation.

Newly-planted soybean plants are emerging from the residue left behind from a prior wheat harvest. This demonstrates the permaculture practices of crop rotation and no-till planting.

Author and restorative development consultant, Storm Cunningham, documented the beginning of what he called "restorative agriculture" in his first book, *The Restoration Economy*. Cunningham defined restorative agriculture as, a technique that rebuilds the quantity and quality of topsoil, while also restoring local biodiversity (especially native pollinators) and watershed function. Carbon sequestration has more recently been added to that definition, to help achieve climate restoration. Restorative agriculture was one of the eight sectors of restorative development industries/disciplines in The Restoration Economy's taxonomy. The other seven sectors were, watershed restoration, ecological restoration, fisheries restoration, brownfield remediation, heritage restoration, infrastructure renewal and catastrophe reconstruction.

2010's onward

Sheep grower, historian, regenerative agriculture consultant and advocate, Charles Massy, published *Call of the Reed Warbler: a new agriculture - a new earth*, based on his PhD studies. The book frames regenerative agriculture as a savior for the earth using case studies.

In 2013,Darren J. Doherty founded Regrarians Ltd., a non-profit promoting regenerative agriculture. The Regrarians organization extends Yeomans' *'Keyline Scale of Permanence'*, adding social and economic perspectives to the original agricultural ones. These perspectives include climate, geography, water, access, forestry, infrastructure, soils, economy and energy.

John Ikerd advocates for the "small" family farm, farmers and sustainability in the US food system. Ikerd is author of *The Essentials of Economic Sustainability*, *Small Farms are Real Farms: Sustaining People through Agriculture* and *Sustainable Capitalism*.

Vermont farmer and farm consultant, Abe Collins, created LandStream to monitor ecosystem performance in regenerative agriculture farms.

Mark Shepard founded New Forest Farms and Forest Agriculture Enterprises in Viola, Wisconsin, and wrote *Restoration Agriculture: Real World Permaculture for Farmers*. He demonstrated how to grow more nutrients per acre than corn and soy, without additional inputs. He does this through a mix of regenerative agriculture practices, nut crops, livestock and keyline design.

Ethan Roland Soloviev and Gregory Landua, cofounded Terra Genesis International (a regenerative agriculture and supply company), and published *Levels of Regenerative Agriculture*. In this paper, they describe a four-fold framework consisting of:

- Functional Regenerative Agriculture: "Humans can do good through their agricultural production".

- Integrative Regenerative Agriculture: "Grow the health and vitality of the whole ecosystem".

- Systemic Regenerative Agriculture: "Farms are woven into an ecosystem of enterprises operating in their bio-region".

- Evolutionary Regenerative Agriculture: "Harmonize with the potential of a place," and "develop a diversity of global and local regenerative producer webs".

Permaculture designer and researcher, Eric Toensmeier, wrote *The Carbon Farming Solution: A*

Global Toolkit of Perennial Crops and Regenerative Agriculture Practices for Climate Change Mitigation and Food Security. Toensmeier claimed that regenerative practices hold the potential to sequester massive amounts of CO_2 into the soil, while providing adaptive and resilient solutions for a changing climate.

Principles and Practices

Regenerative agriculture is guided by a set of principles and practices.

Principles

- Increase soil fertility to nutrient saturation points,
- Work with whole systems, not isolated parts,
- Progressively improve whole agro-ecosystems (soil, water and biodiversity),
- Connect the farm to its larger agro-ecosystem and bio-region,
- Create designs and make holistic decisions that express the value and relationship of farm contributors,
- Express the significance of each person, farm and system,
- Make holistic decisions aimed at specific system changes,
- Ensure and develop equitable and reciprocal relationships among all stakeholders,
- Design for non-linear, multi-capital reciprocity,
- Continually grow and evolve individuals, farms and communities to express their potential,
- Continuously evolve agro-ecological processes and cultures,
- Agriculture influences the world.

Practices

- Permaculture Design,
- Agroforestry,
- Soil Food Web,
- Properly Managed Livestock, Well-managed grazing, Animal Integration, and Holistically Managed Grazing,
- STUN (Sheer, Total and Utter Neglect) Breeding,
- Keyline Subsoiling,
- Conservation farming, No-Till Farming, minimum tillage, and Pasture Cropping,
- Cover crops & multi-species cover crops,

- Organic Annual Cropping and Crop rotations,

- Compost, Compost Tea, animal manures and Thermal Compost,

- Natural sequence farming,

- Grassfed livestock,

- Polyculture and full-time succession planting of multiple and inter-crop plantings,

- Borders planted for pollinator habitat and other beneficial insects,

- Biochar/Terra Preta,

- Ecological Aquaculture,

- Perennial Crops,

- Silvopasture.

GIS APPLICATIONS IN AGRONOMY

Agronomy, an aspect of agriculture, is a spatial activity that represents the backbone of the economy of many nations. This is the result of its noticeable contribution to the employment of labour and the gross domestic product of most developing countries. However, as land is a finite resource, the increase in food production in order to meet an affluent population becomes one of the major issues faced by many developing countries in the world. Hence, the improvement in agronomic practices is inevitable to ensure wise land-use planning and proper management of available resources for Crop cultivation.

With the growing interest in placing site-specific information in a spatial and long-term perspective, precision in agronomic practices would require a technology that can calculate spatial and temporal variations in crop growth with a time scale appropriate for management decisions. Today, advances have been made towards extraordinary digital systems for utilization in soil fertility examination, soil survey and land-use planning, crop production and yield monitoring. Computer programmes, such as geographical information system (GIS), contribute to the speed and efficiency of overall agronomic planning processes.

Most process-based agronomic models examine temporal variations using point data from specific sites, while GIS facilitates storage, manipulations, analysis and visualization of data. They further stated that the interaction of both spatial and temporal issues can be best handled through interfacing agronomic models with geographical information system (GIS).

Importance of GIS to Agronomy

Agronomic activities are spatial and the need to place site-specific information in a spatial and long-term perspective would require special models that can be used to calculate spatial variation in crop growth and monitor variations in trend with a time scale appropriate for guiding decisions. GIS could play a significant role in agronomy at several levels due to the fact that it can be used to study the nutrient status of individual fields to arrive at specific requirements for external application of nutrients. The use of GIS in precision agronomic practices helps to manage the information intensive environment in crop production by combining site-specific (within field) management with computer software modelling for analyses and interpretation of varying inputs and outputs. As opposed to farmers' typical manual adjustment, GIS helps farmers to manage with-in field variable rate application, which results from spatial variation in crop yields within a field. Hence, GIS enhances the assessment and understanding of variations in a field crop. GIS can be used to assemble many layers of information such as soil nutrients, elevation, moisture content and topography to produce a map to show which factors influence crop yield. It was also reported that the yield can then be estimated or used for future reference and the economic inputs and outputs can be calculated based on anticipated yield. This will have a huge potential for saving costs spent on over applied fertilisers that otherwise could have been used on another field.

Applications of GIS in Agronomy

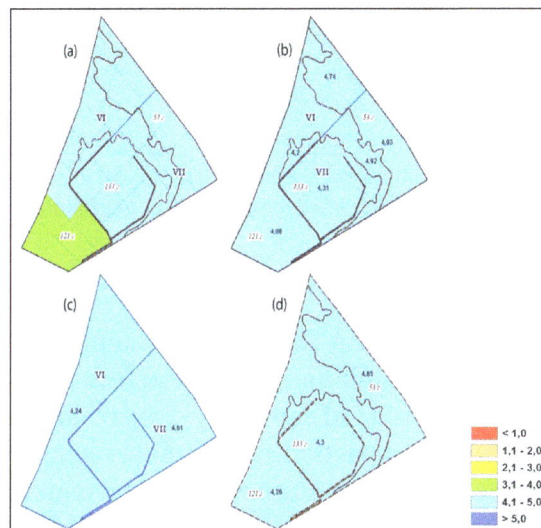

Humus content in the soil: (a) humus content per elementary plots; (b) humus average value per agricultural soil contour per field; (c) average value per field; (d) average value per agricultural soil contours per enterprise.

Applications of GIS have grown from primarily hydrological applications in the mid-1980s to the current wide range of applications in agronomy and natural resource management research. Examples of GIS applications in agronomy and natural resource management research include: Atmospheric modelling, climate change, sensitivity and/or variability studies, characterization and zonation, hydrology, water quality, water pollution, soil science and spatial yield calculation—regional, global and precision farming (spatial yield calculation). Several studies have been reported on the application of GIS on cultivation practices of various crops. The authors reported The application of GIS to fertility management of Soils planted to tea where digitized Maps of the soil pH, potassium, phosphorus and organic matter were prepared using the Arc MAP software. It would be beneficial for tea growers in those locations for calculating fertiliser requirements. It was reported that measures may be required to reduce to a desired level the pH of fields having pH > 5.5. A geodatabase was developed using GIS mapping. This was to provide soil quality monitoring based on data of agrochemical soil survey in order to monitor land cover/soil quality changes between periods of soil survey. In the work of, ArcGIS was employed for mapping soil quality and it was reported that soil data can easily be handled and analysed using ArcGIS because they are spatial in nature. It was also reported in that there was no significant changes in humus and easily hydrolysable nitrogen content within the period between the last two soil agrochemical surveys. A GIS-based decision support system was used to establish potentials and limitations of different soils for crop production, while employed GIS in soil erosion control where the factors and elements affecting erosion were studied by analysing numerical maps of different parts of a basin.

Nitrogen content in the soil: (a) nitrogen content per elementary plots; (b) nitrogen average value per agricultural soil contour per field; (c) average value per field; (d) average value per agricultural soil contours per enterprise.

Operational use of GIS in Precision Farming

The GIS techniques have been used for farm-related assessments for many years at both national and regional scales, respectively. The combination of these techniques and remotely sensed data have been used to aid the assessments of land capability, crop condition and yield range condition, flood and drought soil erosion soil compaction and climate change impacts on regional levels. Also, attempts have been made to assess leaching behaviour for regional scale using a combination of the leaching and chemistry examination (LEACHM) models and GIS database.

At the local level, the number and variety of local agricultural GIS applications have dramatically increased during the past 5 years. Most of the applications are targeted at individual farms. For example, utilized the spatial analysis tools in PC ARC/INFO to perform fully automated conservation program determinations, compliance monitoring and farm planning. This particular application is noteworthy both for its substance and because it illustrates how rapidly the computing resources, user interfaces and database functions in desktop GIS have evolved during the past 5 years. Similarly, determined possible pond sites and estimated rainwater-harvesting potential for a 172-ha farm using GIS.

Most of these field- and subfield-scale applications are connected with precision or site-specific farming, which helps to direct the application of seed, fertiliser, Pesticide and water, within fields in ways that optimize farm returns and minimize chemical inputs and environmental hazards. The use of GIS in precision farming to generate production-based farming system that can be designed to increase long-term, site-specific and whole-farm production efficiency, productivity and profitability was discussed. In addition, it was reported that most site-specific farming systems utilize some combinations of Geographical positioning system (GPS) receivers, continuous yield sensors, remote sensing, geostatistics and variable rate treatment applications with GIS. The reason for combining these advanced technologies is to collect spatially referenced data, perform spatial analysis, make decisions and apply variable rate treatment.

GIS Applications in Agrometeorological Operations

Due to the increasing pressure on land and water resources for crop cultivation, land-use management and forecasting (crop, weather, fire, etc.) have become more essential every day. Hence, GIS is an important tool at the disposal of decision makers. For instance, precipitation and solar radiation are meteorological conditions that can be mapped and monitored to directly assist in the agronomic process to provide advice on the occurrence of drought. It was reported that developed countries use GIS to plan the times and types of agronomic practices, which requires certain information such as soil types, land cover, climatic data and geology, in describing a specific situation in any given location. Each informative layer provides to the operator the possibility to consider its influence on the final outcome.

Operational use of GIS in Agroclimatological and Agroecological Studies

The GIS technology has been shown to synthesize and integrate more data than methods used in the pre-computer era and to shift the design of agroecological and agroclimatological studies towards user-specific classifications. In a study carried out in Zimbabwe, effective rainfall and vegetation for variable interpolation between stations were calculated from rainfall and vegetation data using GIS maps. In addition, seasonal rainfall surfaces were constructed for Zimbabwe using decadal rainfall data while adopting the procedures described. They also generated surfaces showing mean rainfall and annual rainfall anomalies to describe the main rainfall period for Zimbabwe in terms of rainfall variability. This showed the natural regions experiencing considerable spatial variability in terms of mean and inter-seasonal variability of rainfall.

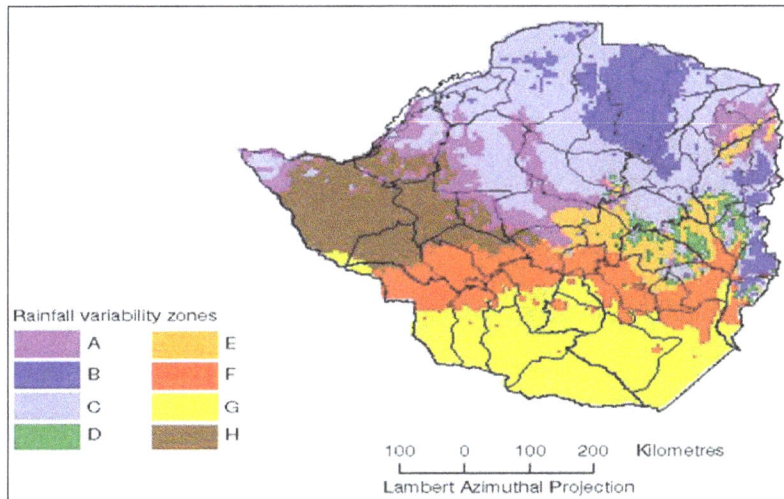

Rainfall variability zones in Zimbabwe.

Use of GIS for Agronomic Characterization and Zonation

The GIS techniques have also been used to characterize agroclimatic diversity and to delineate maize-specific adaptation zones. It was concluded that the emergence of GIS has made it possible to delineate agroclimatic zones with greater precision, especially by allowing many 'layers' of spatially referenced data (including survey data) to be integrated into one digital database.

GIS Application in Soil Survey Studies

Three approaches have been implemented in an attempt to utilize GIS and/or GPS to improve soil attribute predictions at regional scales. The first approach evaluated the use of GIS and/or GPS to improve traditional soil surveys. For example, Long et al. examined the potential of using GPS methods in soil surveys and found these methods to be more efficient than traditional methods of mapping and sufficiently accurate to support positioning/navigating in fields and field digitizing of soil boundaries.

The second approach combined geostatistical modelling with soil survey maps to generate improved soil descriptions. A map that preserved the map unit boundaries and incorporated the spatial variability of the attribute data within the map unit delineations were produced. This was done by combining spatially interpolated (krigged) distributions of measured values with soil map unit delineations within a GIS framework. It was reported by that this approach appeared promising for countries and regions with well-developed soil survey programs.

The third approach neglects the use of traditional soil survey methods and explores the possibilities of integrating GIS, pedology and statistical modelling to improve soil resource inventory. In a study, combined a GIS with an existing soil landscape model to create soil drainage maps. The soil landscape model used multivariate discriminant to predict soil drainage class from parent material, terrain and surface drainage feature variables.

GIS as an Agronomic Land-use Planning Tool

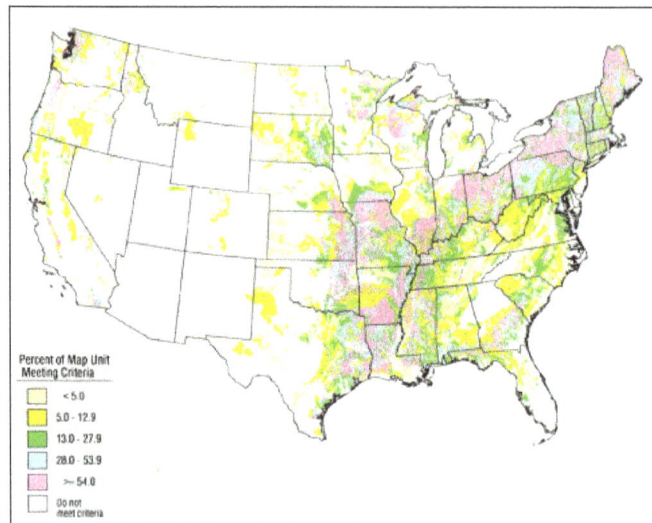

Pictorial view of SPAREC GIS.

Soil survey data and geographic information systems (GIS) are important tools in land-use planning. They reported that the map unit interpretive records (MUIR) were used to create interpretation maps, flooding frequency maps and runoff maps after soil data were added to other data layers and images. Figure shows a flooding frequency map converted from tabular estimates of values in an ArcView GIS. It was explained by that the blue areas are frequently flooded, red areas are occasionally flooded, while the green areas are rarely flooded. They further reported that the soil based-GIS made the decision-making process more accurate, automated and efficient, hence promoting wise land-use planning. In and, it was reported that the soil-based GIS is a dynamic product that serves to convert verbal communication into visual communication while preventing information overload. It was reported that with the GIS, tabular soil information can be georeferenced and easily converted to geographic and interpretive maps, which provides the user with a visual representation of the tabular data. Figure is an example of an interpretive map showing the ratings for site suitability of local roads and streets, where explained that the green areas represent a slight rating, meaning they are the most suitable, while the yellow areas are rated moderate and the red areas are severe areas having the most serious limitations.

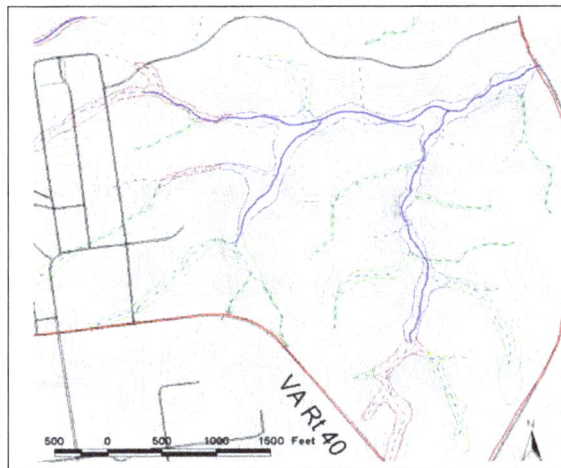

Flooding frequency map.

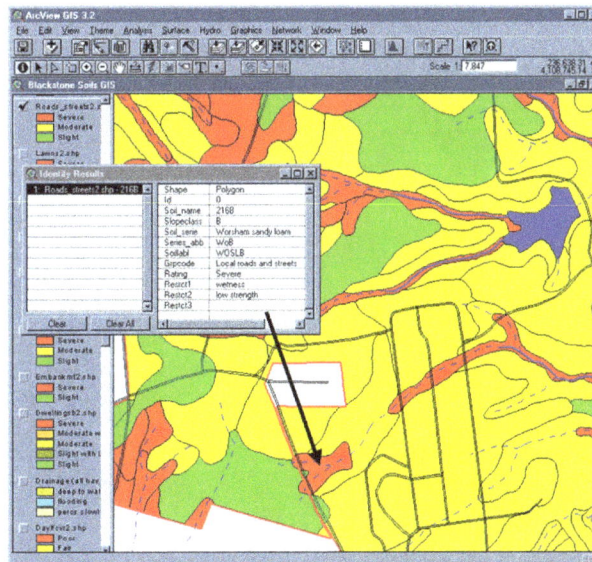

An example of an interpretive map showing ratings for local roads and streets.

Operational use of GIS for Soil Fertility Studies

Soil fertility investigations are necessary to confirm soil fertility status, which is also necessary as a guide for the fertility management practice to adopt. Several methods of soil fertility investigation have been employed in confirming the fertility status of soils. The authors reported that these methods did not ensure the completion of soil fertility investigation within the specified time frame and the required degree of accuracy, as change in soil fertility status over a period of 2 or 3 years makes these methods invalid, thus making it difficult for agronomists to manage soil fertility over large areas. They reported that the application of geospatial technology involving the use of global positioning system (GPS) and geographic information system (GIS) had greatly improved the old traverse techniques.

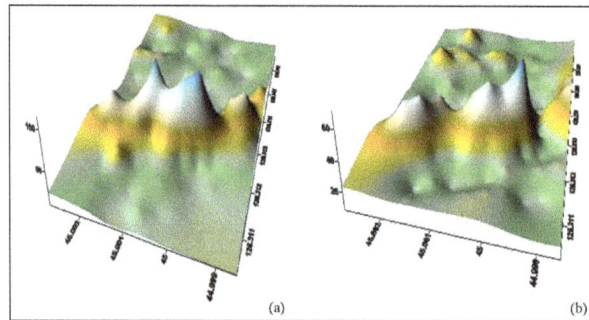

A three-dimensional spatial variability map of available phosphorus for 2003(a) and 2008(b).

In the application of space-time evolution of soil fertility data mining based on visualization, a three-dimensional spatial variation of soil nutrient spatial map for soil available phosphorus was produced by. In a study, evaluated the spatial variation of soil organic carbon, soil water content, $NO_3–N$, $PO_4–P$ (phosphate-phosphorus) and K (potassium) in the 0–15 cm layer of a 3.3 ha field cropped with maize and soya beans. They calculated that as many as 400 randomly selected samples per hectare may be needed to develop an accurate soil $NO_3–N$ map and that an application travelling at 8 km h−1 would need to modulate fertiliser rates every 2.25 s to match nitrogen fertiliser rates to soil $NO_3–N$ requirements.

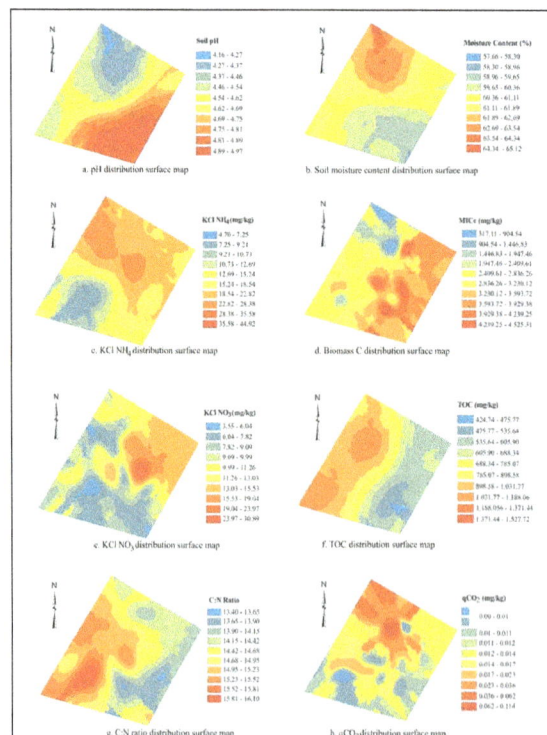

Surface maps showing the distribution of soil fertility indicators.

The authors reported the use of GIS techniques and remote sensing in forest soil fertility studies. GIS could be used to map fertility levels across a farm to serve as basis for the application of farm inputs and also for establishing accurate location of yield data for the production of yield maps for monitoring yield. It was also reported by that periodic review of soil fertility status can be done on digital maps generated with GIS technique. This is due to the fact that the GIS technique uses a digital map which allows the user to view, update, query, analyse and manipulate spatial and tabular data either alone or together, within a few minutes. In assessing the relative efficiency of GIS map-based soil fertility evaluation in relation to traditional soil testing, reported minor variations in available nitrogen content, no variation in available phosphorus and a large difference in available potassium under the two methods of evaluation. They concluded that fertiliser recommendations generated from GIS maps were agronomically as effective as those generated form soil testing.

Parameter	Low/Slightly Acidic		Medium/Acidic		High/Alkaline	
	Soil test	GIS	Soil test	GIS	Soil test	GIS
Available N (g/kg)	8.9	7.8	11	22	0	0
Available P (mg/kg)	100	100	0	0	0	0
Available K (cmol/kg)	44	33	33	67	22	0
pH	5.6	6.7	4.4	3.3	0	0

Comparison of traditional soil test and GIS method of assessing samples (%) that fall under low, medium and high nutrient availability and pH categories.

Treatment	Rice	Potato	Sesame
Farm	60-30-30	300-200-200	Residual
State	80-40-40	200-150-150	80-40-40
Soil test	Variable	Variable	Variable
GIS	Variable	Variable	Variable

Nutrient rates generated from state, field-specific, soil test-based recommendations and GIS.

Spatial Yield Calculation

It was reported that new GIS data layers developed from models were used with some information in various GIS-based application of existing crop yield models. Several studies Showed that these applications can be used to store and process data for decision making with respect to the factors that influence crop cultivation and crop yield in a crop production. For example, the climate surfaces can be used as inputs in genotype-sensitive crop models to assess the risks for specific crop varieties. This was illustrated by those who used GIS and remote sensing technologies with the SOYGRO physiological soya bean growth model to predict the spatial variability of soya bean yields. Continuous yield sensors with a combination of accurate location information obtained using a GPS with the results of a variable flow rate sensor can provide information about the crop performance for a year that can be used to guide the following year's crop management strategies. The examination of spatial patterns of simulated yield improved production estimates and highlighted vulnerable areas during drought.

Agronomic Impact Assessment using GIS

The GIS and environmental models have been combined in many projects to evaluate the impacts of modern agriculture. For instance, used the EPIC-PST crop growth/chemical movement model interfaced with Earthone GIS to evaluate crop yield and nitrate (NO_3-N) movement to surface and ground waters for four soils and nine cropping systems. The authors digitized soil maps using GIS and described how the data can be used with model results to compare the predicted changes in crop yields and nitrogen losses on different soils under water quality protection policies that targets specific soils and/or cropping practices.

References

- Falk, Ben (2013). The Resilient Farm and Homestead: An Innovative Permaculture and Whole Systems Design Approach. Chelsea Green Publishing. P. 280. ISBN 978-1-60358-444-9

- Basic-principles-of-agronomy: blogspot.com, Retrieved 21 April 2019

- Agronomy, agriculture-genera, agriculture-and-horticulture, plants-and-animals: encyclopedia.com, Retrieved 21 May 2019

- Toensmeier, Eric (2016). The Carbon Farming Solution. White River Junction, Vermont: Chelsea Green. ISBN 978-1-60358-571-2

- Role-of-plant-population-for-crop-productivity, basics-of-agriculture: agrihunt.com, Retrieved 2 May 2019

- Tracing the Evolution of Organic / Sustainable Agriculture (TESA1980) | Alternative Farming Systems Information Center| NAL | USDA". Retrieved 2017-03-09

2

Agronomic Practices for Crop Production

The practices that are used to improve the soil quality and the environment, manage crops and enhance water usage are known as agronomic practices. All the diverse methods and practices used in agronomy, such as companion planting, shifting cultivation and crop rotation have been carefully analyzed in this chapter.

Agronomic practices are a vital part of farming systems. These are practices that farmers incorporate to improve soil quality, enhance water usage, manage crops and improve the environment. Agronomic practices focus on better fertilizer management as a way of improving agricultural practices.

Benefits of Agronomic Practices

Proper usage of agronomic practices, decreases input costs in producing farm products. Consequently, the quality and quantity of the yield will increase significantly. The exercises also help the farmer in taking good care of the environment by reducing pollution. Decreasing water usage and proper use of fertilizer also contribute to maintaining the quality of land.

Examples of Agronomic Practices

Agronomic practices incorporate many areas of conservation. In farming any practice that entails conservation is an agronomic practice. Practices such as reducing tillage managing plant population and controlling the use of water are some of the major agronomic practices that almost every farmer has tried. The changes in agronomy might be small, but the results of using the practices are massive. These routines nonetheless have yielded major dividends that farmers have enjoyed.

Practices

Reducing Water Usage

Every farm needs water for the survival of crops and every other use on the farm. Nevertheless, too much usage of water leads to wastage and can sometimes affect the soil. The farm, therefore, needs not too much water such that it's wasted or not too little water such that the soil lacks.

Soil Tillage

The soil is the most vital thing on a farm. Taking good care of it means that the crops will do well and the soil won't lose fertility. In a farm, one needs to practice minimum methods of soil tillage. Minimum soil tillage will increase the water retention capability of the soil as well as maintain its fertility. Maintaining the soil also entails the good use of fertilizer. This will not only take care of the soil but the environment as well.

Manage Plant Populations

Each farmer wants to ensure that they get a bumper harvest. Nevertheless, it is also vital that they ensure their harvest is of quality and not just quantity. Controlling plant population is, therefore, the best way of ensuring crops do well. Managing the population ensures that they are not over crowded such that they compete for nutrients or under crowded such that the input is under-utilized. The plants have to grow in prime growing conditions.

Agronomic practices are efficient and very useful to a farmer. For a commercial farmer, the practices will enable the farmer to get good profits while taking care of the environment. To a domestic farmer, the practices are a cheap method of maintaining your farming projects.

SUBSISTENCE FARMING

Subsistence farming, or subsistence agriculture, is a mode of agriculture in which a plot of land produces only enough food to feed the family or small community working it. All produce grown is intended for consumption purposes as opposed to market sale or trade. Historically and currently a difficult way of life, subsistence farming is considered by many a backward lifestyle that should be transformed into industrialized communities and commercial farming throughout the world in order to overcome problems of poverty and famine. The numerous obstacles that have prevented this to date suggest that a complex array of factors, not only technological but also economic, political, educational, and social, are involved. An alternative perspective, primarily from the feminist voice, maintains that the subsistence lifestyle holds the key to sustainability as human relationships and harmony with the environment have priority over material measures of wealth. Although the poverty suffered by many of those who have never developed beyond subsistence levels of production in farming is something that needs to be overcome, it does appear that the ideas inherent in much of subsistence farming—cooperation, local, ecologically appropriate—are positive attributes that must be preserved in our efforts to improve the lives of all people throughout the world.

Like most farmers in Sub-Saharan Africa, this Cameroonian man cultivates at the subsistence level.

Subsistence farming is a mode of agriculture in which a plot of land produces only enough food to feed those who work it—little or nothing is produced for sale or trade. Depending on climate, soil conditions, agricultural practices and the crops grown, it generally requires between 1,000 and 40,000 square meters (0.25 to 10 acres) per person.

A recognizably harsh way of living, subsistence farmers can experience a rare surplus of produce goods under conditions of good weather which may allow farmers to sell or trade such goods at market. Because such surpluses are rare, subsistence farming does not allow for consistent economic growth and development, the accumulation of capital, or the specialization of labor. Diets of subsistence communities are confined to little else than what is produced by community farmers. Subsistence crops are usually organic due to a lack of finances to buy or trade for industrial inputs such as fertilizer, pesticides or genetically modified seeds.

Subsistence farming, which today exists most commonly throughout areas of Sub-Saharan Africa, Southeast Asia, and parts of South and Central America, is an extension of primitive foraging practiced by early civilizations. Historically, most early farmers engaged in some form of subsistence farming to survive. Within early foraging communities, like hunter-gatherer societies, small communities consumed only what was hunted or gathered by members of the community. As the domestication of certain plants and animals evolved, a more advanced subsistence agricultural society developed in which communities practiced small-scale, low-intensity farming to produce an efficient amount of goods to meet the basic consumption needs of the community.

Historically, successful subsistence farming systems often shared similar structural traits. These included equal access to land plots for community members as well as a minimum expenditure of agricultural labor to produce subsistence amounts of food. Over time, the loss of such freedoms forced many subsistence farmers to abandon their traditional ways. In early twentieth-century Kenya, a lack of land access due to the commercialization of certain farmland plots by British colonists forced Kenyan communities toward commercial farming. Consistent surpluses, like those experienced by nineteenth century South Africa and sixteenth century Japan, also encouraged commercialized production and allowed farmers to expend more amounts of agricultural labor on certain produce goods that were strictly intended for trade.

Though forms of subsistence farming are believed to have been practiced by most early civilizations worldwide, over time, as population densities rose and intensive farming methods developed, the movement toward commercial farming and industrialization became more prominent. For countries like Botswana, Bolivia, Rwanda, Sierra Leone, Zambia, Mexico, and Vietnam, however, subsistence farming continues to be a way of life far into the twenty-first century.

Techniques

In the absence of technology, the area of land that a farmer can cultivate each season is limited by factors such as available tools and the quality of the soil. Tools used by subsistence farmers are often primitive. Most farmers do not have access to large domesticated work animals, and therefore clear, toil, and harvest their goods using pointed sticks, hoes, or by hand.

Techniques of subsistence farming include "slash and burn" clearing in which farmers clear plots of farmland by cutting down all brush, allowing the debris to dry, and later burning the fallen refuse. This works to clear the field for cultivation, while the leftover ash serves as a natural fertilizer. This type of clearing technique is often employed by subtropical communities throughout lush areas of South and Central America, and parts of Indonesia.

If the land does not produce a surplus, due to the fertility of the soil, climate conditions, tools and techniques, or available crop types, the farmer can do no more than hope to subsist on it. Under these conditions, subsequent years with poor harvests often result in food scarcity and famine.

Not all subsistence farmers have access to as much land as they can cultivate. Many times, socio-economic conditions prevent an expansion of farming plots and any increase in produce levels. If inheritance traditions require that a plot be split among an owner's children upon the owner's death, plot sizes steadily decrease.

Industrial Intervention

Many techniques have been attempted, with varying degrees of success, to help subsistence farmers to produce consistent surpluses so that small underdeveloped communities can begin the path toward commercial farming, and economic development.

Education about modern agricultural techniques has proven to have limited success in areas practicing subsistence farming. Since subsistence communities often lack the basic infrastructure for industrial growth, a second approach to education has been to provide community farmers with non-agricultural marketable skills. Under this approach, subsistence farmers are given an opportunity to leave the subsistence community to seek employment in an area where greater resources are available. This technique has been met with marginal success as it often ignores the human desire to stay within one's own community.

Attention has also been given to developing underutilized crops, particularly in areas of Africa and South-East Asia. Genetically modified crops, such as golden rice, have also been used to improve productivity within subsistence communities. Such crops are proven to have higher nutrient content or disease resistance than natural varieties, and represent an increase in farming efficiency. This technique has been highly successful in some parts of the world, although long-term ecological and epidemiological effects of these crops are often poorly understood.

Proper irrigation techniques can also dramatically improve the productivity of subsistence farmland and have been introduced to certain rural communities in hopes of promoting output surpluses. Traditional irrigation methods, if in place, have been shown to be extremely labor-intensive, wasteful of water, and may require a community-wide infrastructure which is difficult to implement. A variety of programs have helped to introduce new types of irrigation equipment available which are both inexpensive and water-efficient. Many subsistence farmers, however, are often unaware of such technologies, are unable to afford them, or have difficulty marketing their crops after investing in irrigation equipment.

Microloans, or government loans of small sums of money, have also been shown to enable farmers to purchase equipment or draft animals. Alternatively, microloans may enable farmers to find non-agricultural occupations within their communities.

Obstacles to Industrial Development

Peruvian economist Hernando de Soto has argued that one obstacle to industrial development is that subsistence farmers cannot convert their work into capital which could ultimately be used to start new businesses and trigger industrialization. De Soto has argued that these obstacles exist often because subsistence farmers do not have clear ownership titles to the land which they work and to the crops which they produce.

In addition to the problems presented by undefined property rights, monetary demands on industrial producers, like produce taxes, often dissuade subsistence farmers from entering the commercial farming sector. Moreover, the marginal benefit of surplus production is limited, and any extra effort to increase production is poorly rewarded.

Subsistence farmers in underdeveloped countries often lack equal access to trade markets. Despite attempts to specialize in the production and distribution of certain crops, many subsistence communities still lack access to open market systems in which the sale or trade of such goods are possible. In addition, educational studies have shown certain industrial growth techniques to depend on various infrastructures, climates, or resources that are not available in all communities relying on subsistence farming. In this way, subsistence farming may represent the only way many deeply rural communities can survive.

Subsistence Farming and the Modern World

Despite its difficulties, subsistence farming remains a part of the modern world today. For many underdeveloped nations, subsistence farming represents the only option to prevent starvation and famine.

Subsistence farming has been argued to be economically efficient within various subtropical regions of Columbia and Papua New Guinea. Under these subtropical conditions, rainfall levels are often high and various crops can be produced year round. Due to these conditions, production levels often prove adequate enough to provide for small subsistence farming communities.

Subsistence Farming in Zambia

This argument does not hold for many Sub-Saharan regions of Africa, where poverty and famine levels are some of the highest in the world. One reason why subsistence farming systems have failed through-out the Sub-Saharan region are increasing trends in population growth that are not met with an equal increase in the production of agricultural output. Other reasons include unusually harsh climate conditions, widespread disease among plants and animals, and a lack of efficient institutional structures.

In parts of rural Zambia, much of the current population relies on subsistence farming to survive. As irrigation systems are few, most Zambians must rely on seasonal rains to ensure crop production. In 1995, Zambia underwent a severe drought which vastly diminished production levels throughout traditional farming communities. Similar impoverishment has been observed throughout parts of the Amazon Basin of Brazil and the Indonesian islands of Sumatra and Borneo, which also rely heavily on subsistence farming and production.

Subsistence farming in Zambia.

Many developmental economists have argued against the use of subsistence farming and instead promote commercial farming and economic industrialization as the solution to worldwide hunger. Economist Ronald E. Seavoy, author of Subsistence and Economic Development, argued that subsistence farming is to blame for high levels of poverty and increasing instances of famine, recommending the transformation of subsistence agriculture into commercial agriculture which would ultimately promote economic development among economically underdeveloped nations.

Attempts have continued to be made to move in this direction away from subsistence farming. In central Uganda commercial farming has been promoted to alleviate high poverty levels throughout Ugandan subsistence farming communities. Restructuring the production output of the people and identifying a potential market for free trade, are key to successful small-scale industrialization, thereby improving rural living conditions and diminishing poverty rates.

An alternative viewpoint, particularly promoted by women often called "ecofeminists," reflects the need to understand sustainable economies. Those such as Maria Mies and Vandana Shiva have argued that the free market capitalist system is inherently unsustainable in the long run, since it exploits various population groups and the environment. Instead, they argue that the "catch-up" model of economic development, assuming that western-style progress is possible and optimal for all, be replaced with a more ecologically sensitive approach, valuing harmony with nature and the goals of happiness, quality of life, and human dignity over the accumulation of wealth. They explain subsistence as empowerment for all, based on people's strengths and their cooperation with nature and each other.

COMPANION PLANTING

Companion planting in gardening and agriculture is the planting of different crops in proximity for any of a number of different reasons, including pest control, pollination, providing habitat for beneficial insects, maximizing use of space, and to otherwise increase crop productivity. Companion planting is a form of polyculture.

Companion planting is used by farmers and gardeners in both industrialized and developing countries for many reasons. Many of the modern principles of companion planting were present many centuries ago in cottage gardens in England and forest gardens in Asia, and thousands of years ago in Mesoamerica.

Mechanisms

Companion planting can operate through a variety of mechanisms, which may sometimes be combined.

Provision of Nutrients

Legumes such as clover provide nitrogen compounds to other plants such as grasses by fixing nitrogen from the air with symbiotic bacteria in their root nodules.

Dandelions have long taproots that bring nutrients from deep within the soil to near the surface, benefitting neighboring plants that are shallower-rooted.

Trap Cropping

Trap cropping uses alternative plants to attract pests away from a main crop. For example, nasturtium (*Tropaeolum majus*) is a food plant of some caterpillars which feed primarily on members of the cabbage family (brassicas); some gardeners claim that planting them around brassicas protects the food crops from damage, as eggs of the pests are preferentially laid on the nasturtium. However, while many trap crops have successfully diverted pests off of focal crops in small scale greenhouse, garden and field experiments, only a small portion of these plants have been shown to reduce pest damage at larger commercial scales.

Host-finding Disruption

Recent studies on host-plant finding have shown that flying pests are far less successful if their host-plants are surrounded by any other plant or even "decoy-plants" made of green plastic, cardboard, or any other green material.

The host-plant finding process occurs in phases:

- The first phase is stimulation by odours characteristic to the host-plant. This induces the insect to try to land on the plant it seeks. But insects avoid landing on brown (bare) soil. So if only the host-plant is present, the insects will quasi-systematically find it by simply landing on the only green thing around. This is called (from the point of view of the insect) "appropriate landing". When it does an "inappropriate landing", it flies off to any other nearby patch of green. It eventually leaves the area if there are too many 'inappropriate' landings.

- The second phase of host-plant finding is for the insect to make short flights from leaf to leaf to assess the plant's overall suitability. The number of leaf-to-leaf flights varies according to the insect species and to the host-plant stimulus received from each leaf. The insect must accumulate sufficient stimuli from the host-plant to lay eggs; so it must make a certain number of consecutive 'appropriate' landings. Hence if it makes an 'inappropriate landing', the assessment of that plant is negative, and the insect must start the process anew.

Thus it was shown that clover used as a ground cover had the same disruptive effect on eight pest species from four different insect orders. An experiment showed that 36% of cabbage root flies laid eggs beside cabbages growing in bare soil (which resulted in no crop), compared to only 7% beside cabbages growing in clover (which allowed a good crop). Simple decoys made of green cardboard also disrupted appropriate landings just as well as did the live ground cover.

Pest Suppression

Some companion plants help prevent pest insects or pathogenic fungi from damaging the crop, through chemical means. For example, the smell of the foliage of marigolds is claimed to deter aphids from feeding on neighbouring plants.

Predator Recruitment

Companion plants that produce copious nectar or pollen in a vegetable garden (insectary plants) may help encourage higher populations of beneficial insects that control pests, as some beneficial predatory insects only consume pests in their larval form and are nectar or pollen feeders in their adult form. For instance, marigolds with simple flowers attract nectar-feeding adult hoverflies, the larvae of which are predators of aphids.

Protective Shelter

Some crops are grown under the protective shelter of different kinds of plant, whether as wind breaks or for shade. For example, shade-grown coffee, especially *Coffea arabica*, has traditionally been grown in light shade created by scattered trees with a thin canopy, allowing light through to the coffee bushes but protecting them from overheating. Suitable Asian trees include *Erythrina subumbrans* (tton tong or dadap), *Gliricidia sepium* (khae falang), *Cassia siamea* (khi lek), *Melia azedarach* (khao dao sang), and *Paulownia tomentosa*, a useful timber tree.

Shade-grown coffee plantation in Costa Rica. The red trees in the background provide shade; those in the foreground have been pruned to allow full exposure to the sun.

Systems

Systems in use or being trialled include:

- Square foot gardening attempts to protect plants from many normal gardening problems, such as weed infestation, by packing them as closely together as possible, which is facilitated by using companion plants, which can be closer together than normal.

- Forest gardening, where companion plants are intermingled to create an actual ecosystem, emulates the interaction of up to seven levels of plants in a forest or woodland.

- Organic gardening makes frequent use of companion planting, since many other means of fertilizing, weed reduction and pest control are forbidden.

SHIFTING CULTIVATION

Shifting cultivation is an agricultural system in which plots of land are cultivated temporarily, then abandoned and allowed to revert to their natural vegetation while the cultivator moves on to another plot. The period of cultivation is usually terminated when the soil shows signs of exhaustion or, more commonly, when the field is overrun by weeds. The length of time that a field is cultivated is usually shorter than the period over which the land is allowed to regenerate by lying fallow. This technique is often used in LEDCs (Less Economically Developed Countries) or LICs (Low Income Countries). In some areas, cultivators use a practice of slash-and-burn as one element of their farming cycle. Others employ land clearing without any burning, and some cultivators are purely migratory and do not use any cyclical method on a given plot. Sometimes no slashing at all is needed where regrowth is purely of grasses, an outcome not uncommon when soils are near exhaustion and need to lie fallow. In shifting agriculture, after two or three years of producing vegetable and grain crops on cleared land, the migrants abandon it for another plot. Land is often cleared by slash-and-burn methods—trees, bushes and forests are cleared by slashing, and the remaining vegetation is burnt. The ashes add potash to the soil. Then the seeds are sown after the rains.

Political Ecology

Shifting cultivation is a form of agriculture or a cultivation system, in which, at any particular point in time, a minority of 'fields' are in cultivation and a majority are in various stages of natural re-growth. Over time, fields are cultivated for a relatively short time, and allowed to recover, or are fallowed, for a relatively long time. Eventually a previously cultivated field will be cleared of the natural vegetation and planted in crops again. Fields in established and stable shifting cultivation systems are cultivated and fallowed cyclically. This type of farming is called jhumming in India.

Fallow fields are not unproductive. During the fallow period, shifting cultivators use the successive vegetation species widely for timber for fencing and construction, firewood, thatching, ropes, clothing, tools, carrying devices and medicines. It is common for fruit and nut trees to be planted in fallow fields to the extent that parts of some fallows are in fact orchards. Soil-enhancing shrub or tree species may be planted or protected from slashing or burning in fallows. Many of these species have been shown to fix nitrogen. Fallows commonly contain plants that attract birds and animals and are important for hunting. But perhaps most importantly, tree fallows protect soil against physical erosion and draw nutrients to the surface from deep in the soil profile.

The relationship between the time the land is cultivated and the time it is fallowed are critical to the stability of shifting cultivation systems. These parameters determine whether or not the shifting cultivation system as a whole suffers a net loss of nutrients over time. A system in which there is a net loss of nutrients with each cycle will eventually lead to a degradation of resources unless actions are taken to arrest the losses. In some cases soil can be irreversibly exhausted (including erosion as well as nutrient loss) in less than a decade.

The longer a field is cropped, the greater the loss of soil organic matter, cation-exchange-capacity and in nitrogen and phosphorus, the greater the increase in acidity, the more likely soil porosity and infiltration capacity is reduced and the greater the loss of seeds of naturally occurring plant species from soil seed banks. In a stable shifting cultivation system, the fallow is long enough for the natural vegetation to recover to the state that it was in before it was cleared, and for the soil to recover to the condition it was in before cropping began. During fallow periods soil temperatures are lower, wind and water erosion is much reduced, nutrient cycling becomes closed again, nutrients are extracted from the subsoil, soil fauna decreases, acidity is reduced, soil structure, texture and moisture characteristics improve and seed banks are replenished.

The secondary forests created by shifting cultivation are commonly richer in plant and animal resources useful to humans than primary forests, even though they are much less bio-diverse. Shifting cultivators view the forest as an agricultural landscape of fields at various stages in a regular cycle. People unused to living in forests cannot see the fields for the trees. Rather they perceive an apparently chaotic landscape in which trees are cut and burned randomly and so they characterise shifting cultivation as ephemeral or 'pre-agricultural', as 'primitive' and as a stage to be progressed beyond. Shifting agriculture is none of these things. Stable shifting cultivation systems are highly variable, closely adapted to micro-environments and are carefully managed by farmers during both the cropping and fallow stages. Shifting cultivators may possess a highly developed knowledge and understanding of their local environments and of the crops and native plant species they exploit. Complex and highly adaptive land tenure systems sometimes exist under shifting cultivation. Introduced crops for food and as cash have been skillfully integrated into some shifting cultivation systems. Its disadvantages include the high initial cost, as manual labour is required.

Simple Societies and Environmental Change

Shifting cultivation in Indonesia. A new crop is sprouting through the burnt soil.

A growing body of palynological evidence finds that simple human societies brought about extensive changes to their environments before the establishment of any sort of state, feudal or capitalist, and before the development of large scale mining, smelting or shipbuilding industries. In these societies agriculture was the driving force in the economy and shifting cultivation was the most common type of agriculture practiced. By examining the relationships between social and economic change and agricultural change in these societies, insights can be gained on contemporary social and economic change and global environment change, and the place of shifting cultivation in those relationship.

As early as 1930 questions about relationships between the rise and fall of the Mayan civilization of the Yucatán Peninsula and shifting cultivation were raised and continue to be debated today. Archaeological evidence suggests the development of Mayan society and economy began around 250 AD. A mere 700 years later it reached its apogee, by which time the population may have reached 2,000,000 people. There followed a precipitous decline that left the great cities and ceremonial centres vacant and overgrown with jungle vegetation. The causes of this decline are uncertain; but warfare and the exhaustion of agricultural land are commonly cited. More recent work suggests the Maya may have, in suitable places, developed irrigation systems and more intensive agricultural practices.

Similar paths appear to have been followed by Polynesian settlers in New Zealand and the Pacific Islands, who within 500 years of their arrival around 1100 AD turned substantial areas from forest into scrub and fern and in the process caused the elimination of numerous species of birds and animals. In the restricted environments of the Pacific islands, including Fiji and Hawaii, early extensive erosion and change of vegetation is presumed to have been caused by shifting cultivation on slopes. Soils washed from slopes were deposited in valley bottoms as a rich, swampy alluvium. These new environments were then exploited to develop intensive, irrigated fields. The change from shifting cultivation to intensive irrigated fields occurred in association with a rapid growth in population and the development of elaborate and highly stratified chiefdoms. In the larger, temperate latitude, islands of New Zealand the presumed course of events took a different path. There the stimulus for population growth was the hunting of large birds to extinction, during which time forests in drier areas were destroyed by burning, followed the development of intensive agriculture in favorable environments, based mainly on sweet potato (Ipomoea batatas) and a reliance on the gathering of two main wild plant species in less favorable environments. These changes, as in the smaller islands, were accompanied by population growth, the competition for the occupation of the best environments, complexity in social organization, and endemic warfare.

The record of humanly induced changes in environments is longer in New Guinea than in most places. Agricultural activities probably began 5,000 to 9,000 years ago. However, the most spectacular changes, in both societies and environments, are believed to have occurred in the central highlands of the island within the last 1,000 years, in association with the introduction of a crop new to New Guinea, the sweet potato. One of the most striking signals of the relatively recent intensification of agriculture is the sudden increase in sedimentation rates in small lakes.

The root question posed by these and the numerous other examples that could be cited of simple societies that have intensified their agricultural systems in association with increases in population and social complexity is not whether or how shifting cultivation was responsible for the extensive changes to landscapes and environments. Rather it is why simple societies of shifting cultivators in the tropical forest of Yucatán, or the highlands of New Guinea, began to grow in numbers and to develop stratified and sometimes complex social hierarchies?

At first sight, the greatest stimulus to the intensification of a shifting cultivation system is a growth in population. If no other changes occur within the system, for each extra person to be fed from the system, a small extra amount of land must be cultivated. The total amount of land available is the land being presently cropped and all of the land in fallow. If the area occupied by the system is not expanded into previously unused land, then either the cropping period must be extended or the fallow period shortened.

At least two problems exist with the population growth hypothesis. First, population growth in most pre-industrial shifting cultivator societies has been shown to be very low over the long term. Second, no human societies are known where people work only to eat. People engage in social relations with each other and agricultural produce is used in the conduct of these relationships.

These relationships are the focus of two attempts to understand the nexus between human societies and their environments, one an explanation of a particular situation and the other a general exploration of the problem.

Feedback Loops

In a study of the Duna in the Southern Highlands of New Guinea, a group in the process of moving from shifting cultivation into permanent field agriculture post sweet potato, Modjeska argued for the development of two "self amplifying feed back loops" of ecological and social causation. The trigger to the changes were very slow population growth and the slow expansion of agriculture to meet the demands of this growth. This set in motion the first feedback loop, the "use-value" loop. As more forest was cleared there was a decline in wild food resources and protein produced from hunting, which was substituted for by an increase in domestic pig raising. An increase in domestic pigs required a further expansion in agriculture. The greater protein available from the larger number of pigs increased human fertility and survival rates and resulted in faster population growth.

The outcome of the operation of the two loops, one bringing about ecological change and the other social and economic change, is an expanding and intensifying agricultural system, the conversion of forest to grassland, a population growing at an increasing rate and expanding geographically and a society that is increasing in complexity and stratification.

Resources are Cultural Appraisals

The second attempt to explain the relationships between simple agricultural societies and their environments is that of Ellen. Ellen does not attempt to separate use-values from social production. He argues that almost all of the materials required by humans to live (with perhaps the exception of air) are obtained through social relations of production and that these relations proliferate and are modified in numerous ways. The values that humans attribute to items produced from the environment arise out of cultural arrangements and not from the objects themselves, a restatement of Carl Sauer's dictum that "resources are cultural appraisals". Humans frequently translate actual objects into culturally conceived forms, an example being the translation by the Duna of the pig into an item of compensation and redemption. As a result, two fundamental processes underlie the ecology of human social systems: First, the obtaining of materials from the environment and their alteration and circulation through social relations, and second, giving the material a value which will affect how important it is to obtain it, circulate it or alter it. Environmental pressures are thus mediated through social relations.

Transitions in ecological systems and in social systems do not proceed at the same rate. The rate of phylogenetic change is determined mainly by natural selection and partly by human interference and adaptation, such as for example, the domestication of a wild species. Humans however have the ability to learn and to communicate their knowledge to each other and across generations. If most social systems have the tendency to increase in complexity they will, sooner or later, come into conflict with, or into "contradiction" with their environments. What happens around the point of "contradiction" will determine the extent of the environmental degradation that will occur. Of particular importance is the ability of the society to change, to invent or to innovate technologically and sociologically, in order to overcome the "contradiction" without incurring continuing environmental degradation, or social disintegration.

An economic study of what occurs at the points of conflict with specific reference to shifting cultivation is that of Esther Boserup. Boserup argues that low intensity farming, extensive shifting cultivation for example, has lower labor costs than more intensive farming systems. This assertion remains controversial. She also argues that given a choice, a human group will always choose the technique which has the lowest absolute labor cost rather than the highest yield. But at the point of conflict, yields will have become unsatisfactory. Boserup argues, contra Malthus, that rather than population always overwhelming resources, that humans will invent a new agricultural technique or adopt an existing innovation that will boost yields and that is adapted to the new environmental conditions created by the degradation which has occurred already, even though they will pay for the increases in higher labor costs. Examples of such changes are the adoption of new higher yielding crops, the exchanging of a digging stick for a hoe, or a hoe for a plough, or the development of irrigation systems. The controversy over Boserup's proposal is in part over whether intensive systems are more costly in labor terms, and whether humans will bring about change in their agricultural systems before environmental degradation forces them to.

Contemporary World and Global Environmental Change

Rio Xingu, Brazil.

The estimated rate of deforestation in Southeast Asia in 1990 was 34,000 km² per year. In Indonesia alone it was estimated 13,100 km² per year were being lost, 3,680 km² per year from Sumatra and 3,770 km² from Kalimantan, of which 1,440 km² were due to the fires of 1982 to 1983. Since those estimates were made huge fires have ravaged Indonesian forests during the 1997 to 1998 El Niño associated drought.

Santa Cruz, Bolivia.

Shifting cultivation was assessed by the FAO to be one of the causes of deforestation while logging was not. The apparent discrimination against shifting cultivators caused a confrontation between FAO and environmental groups, who saw the FAO supporting commercial logging interests against the rights of indigenous people. Other independent studies of the problem note that despite lack of government control over forests and the dominance of a political elite in the logging industry, the causes of deforestation are more complex. The loggers have provided paid employment to former subsistence farmers. One of the outcomes of cash incomes has been rapid population growth among indigenous groups of former shifting cultivators that has placed pressure on their traditional long fallow farming systems. Many farmers have taken advantage of the improved road access to urban areas by planting cash crops, such as rubber or pepper as noted above. Increased cash incomes often are spent on chain saws, which have enabled larger areas to be cleared for cultivation. Fallow periods have been reduced and cropping periods extended. Serious poverty elsewhere in the country has brought thousands of land-hungry settlers into the cut-over forests along the logging roads. The settlers practice what appears to be shifting cultivation but which is in fact a one-cycle slash and burn followed by continuous cropping, with no intention to long fallow. Clearing of trees and the permanent cultivation of fragile soils in a tropical environment with little attempt to replace lost nutrients may cause rapid degradation of the fragile soils.

The loss of forest in Indonesia, Thailand, and the Philippines during the 1990s was preceded by major ecosystem disruptions in Vietnam, Laos and Cambodia in the 1970s and 1980s caused by warfare. Forests were sprayed with defoliants, thousands of rural forest dwelling people uproots from their homes and moved and roads driven into previously isolated areas. The loss of the tropical forests of Southeast Asia is the particular outcome of the general possible outcomes described by Ellen when small local ecological and social systems become part of larger system. When the previous relatively stable ecological relationships are destabilized, degradation can occur rapidly. Similar descriptions of the loss of forest and destruction of fragile ecosystems could be provided from the Amazon Basin, by large scale state sponsored colonization forest land or from the Central Africa where what endemic armed conflict is destabilizing rural settlement and farming communities on a massive scale.

Comparison with other Ecological Phenomena

In the tropical developing world, shifting cultivation in its many diverse forms, remains a pervasive practice. Shifting cultivation was one of the very first forms of agriculture practiced by humans and its survival into the modern world suggests that it is a flexible and highly adaptive means of production. However, it is also a grossly misunderstood practice. Many casual observers cannot see past the clearing and burning of standing forest and do not perceive often ecologically stable cycles of cropping and fallowing. Nevertheless, shifting cultivation systems are particularly susceptible to rapid increases in population and to economic and social change in the larger world around them. The blame for the destruction of forest resources is often laid on shifting cultivators. But the forces bringing about the rapid loss of tropical forests at the end of the 20th century are the same forces that led to the destruction of the forests of Europe, urbanization, industrialization, increased affluence, populational growth and geographical expansion and the application the latest technology to extract ever more resources from the environment in pursuit of wealth and political power by competing groups. However we must know that those who practice Agriculture are at the receiving end of the social stratum.

Studies of small, isolated and pre-capitalist groups and their relationships with their environments suggests that the roots of the contemporary problem lie deep in human behavioral patterns, for even in these simple societies, competition and conflict can be identified as the main force driving them into contradiction with their environments.

CROP ROTATION

Crop rotation is the successive cultivation of different crops in a specified order on the same fields, in contrast to a one-crop system or to haphazard crop successions.

Throughout human history, wherever food crops have been produced, some kind of rotation cropping appears to have been practiced. One system in central Africa employs a 36-year rotation; a single crop of finger millet is produced after a 35-year growth of woody shrubs and trees has been cut and burned. In the major food-producing regions of the world, various rotations of much shorter length are widely used. Some of them are designed for the highest immediate returns, without much regard for the continuing usefulness of the basic resources. Others are planned for high continuing returns with protected resources. The underlying principles for planning effective cropping systems began to emerge in the middle years of the 19th century.

Early experiments, such as those at the Rothamsted experimental station in England in the mid-19th century, pointed to the usefulness of selecting rotation crops from three classifications: Cultivated row, close-growing grains, and sod-forming, or rest, crops. Such a classification provides a ratio basis for balancing crops in the interest of continuing soil protection and production economy. It is sufficiently flexible for adjusting crops to many situations, for making changes when needed, and for including go-between crops as cover and green manures.

A simple rotation would be one crop from each group with a 1:1:1 ratio. The first number in a rotation ratio refers to cultivated row crops, the second to close-growing grains, and the third to sod-forming, or rest, crops. Such a ratio signifies the need for three fields and three years to produce each crop annually. This requirement would be satisfied with a rotation of corn, oats, and clover or of potatoes, wheat, and clover-timothy. Rotations for any number of fields and crop relationships can be described in this manner. In general, most rotations are confined to time limits of eight years or less.

The acreage devoted to sod-forming, or rest, crops should be expanded at the expense of row crops on soils of increasing slopes and declining fertility. This will provide better vegetative covering to protect sloping land from excessive erosion and supply organic matter for improving soil productivity on both sloping and level lands. With lessening slope and increasing fertility, the row crops may be expanded, but this should not be done with too much reduction in the sod-forming crops. The differing effects of crops on soils and on each other and in reactions to insect pests, diseases, and weeds require carefully planned sequences.

Broadly speaking, cropping systems should be planned around the use of deep-rooting legumes. If too little use is made of them, productivity will decline; if too much land is devoted to them, wastes may occur and other useful crops will be displaced. Rotations depending wholly on green-manure legumes should be confined to the more level and fertile lands. It is desirable to include legumes alone or in mixtures with nonlegume sod-forming crops as a regular crop in many field rotations. In general, this should occur about once in each four-year period. Short rotations are not likely to provide the best crop balances, and long rotations on a larger number of fields may introduce complications. With a moderate number of fields, additional flexibility can be provided by split cropping on some fields.

The usefulness of individual field crops is affected by regional differences in climate and soil. A major crop in one region may have little or no value in another. In each region, however, there are usually row, grain, and sod, or rest, crops that can be brought together into effective cropping systems.

In addition to the many beneficial effects on soils and crops, well-planned crop rotations also provide the business aspects of farming with advantages. Labour, power, and equipment can be handled with more efficiency; weather and market risks can be reduced; livestock requirements can be met more easily; and the farm can be a more effective year-round enterprise.

Advantages of Crop Rotation

Crop rotation is the practice of growing a series of different types of crops in the same area in sequential seasons. Crop rotation gives various nutrients to the soil. A traditional element of crop rotation is the replenishment of nitrogen through the use of green manure in sequence with cereals and other crops. Crop rotation also mitigates the build-up of pathogens and pests that often occurs when one species is continuously cropped, and can also improve soil structure and fertility by alternating deep-rooted and shallow-rooted plants.

Crop rotation has increased in the south in the last 10 years due to the changing tides of the ever changing grain price. With the increase in corn acres across the south, as well as the increase in irrigation, we have seen a steady increase in yields. There are many studies showing yield increases of 10 to 15 percent in soybeans and corn when rotation is utilized.

Rotations also help with a reduction in nematodes, weeds and diseases. Northern Leaf Blight is a good example of a disease that has increased over the last several years, and can be reduced by rotating corn and soybeans.

Understanding the relationship between nitrogen (N) and crop rotation is very important when making N management decisions. There are several benefits to using crop rotation, including improved nutrient cycling, soil tilth, and soil physical properties; and enhanced weed control. Crop rotation also may influence the rate of N mineralization or the conversion of organic N to mineral N by modifying soil moisture, soil temperature, pH, plant residue, and tillage practices.

The incremental increase in N use over the past five decades, due to emphasis on maximizing yield, has led to a subsequent increase in N in the soil profile of some agricultural fields. Therefore, the influence of agricultural practices on water quality has prompted studies to develop best management practices to optimize the use of fertilizer N and reduce N loss to surface and groundwater. Crop rotation can play a major role in minimizing the potential risk of nitrate leaching to surface and groundwater by enhancing soil N availability, reducing the amount of N fertilizer applied, and minimizing the potential risk of N leaching.

Research on the impact of long-term crop rotation on soil N availability shows that planting alfalfa, corn, oat, and soybean significantly increased the mineralized net N in soil compared with planting continuous corn. Because soil N mineralization can effect yield, crop rotation thus can be used as a management system to enhance the soil nutrient pool, thereby reducing the fertilizer N input and minimizing the risk of leaching of excess N during wet weather.

A combination of conservation tillage practices and crop rotation has been shown to be very effective in improving soil physical properties. Long-term studies in the Midwest indicate that corn-soybean rotation improves yield potential of no-till compared with continuous corn. The reduction in yield of continuous corn in no-till is attributed to low soil temperature during seed germination, which is evident on poorly drained soils under no-till. Studies show that the poor performance of no-till corn following corn is more likely due to the previous crop than to surface residue conditions preventing early-season warming and drying of soils.

The use of a legume cover in crop rotation can provide a substantial amount of N to a succeeding crop. Research has indicated that seeding rates for legumes can be reduced by approximately one-third of that recommended for forage production when used as cover crops without sacrificing biomass or N accumulation. Also, the type of crop grown in the previous year can impact the efficiency of conservation tillage, especially for no-till systems, due to the kind and amount of crop residue from the previous crop.

ORGANIC FARMING

Organic farming is an alternative agricultural system which originated early in the 20th century in reaction to rapidly changing farming practices. Certified organic agriculture accounts for 70 million hectares globally, with over half of that total in Australia. Organic farming continues to be developed by various organic agriculture organizations today. It is defined by the use of fertilizers of organic origin such as compost manure, green manure, and bone meal and places emphasis on techniques such as crop rotation and companion planting. Biological pest control, mixed cropping and the fostering of insect predators are encouraged. In general, organic standards are designed to allow the use of naturally occurring substances while prohibiting or strictly limiting synthetic substances. For instance, naturally occurring pesticides such as pyrethrin and rotenone are permitted, while synthetic fertilizers and pesticides are generally prohibited. Synthetic substances that are allowed include, for example, copper sulfate, elemental sulfur and Ivermectin. Genetically modified organisms, nano-materials, human sewage sludge, plant growth regulators, hormones, and antibiotic use in livestock husbandry are prohibited. Reasons for advocation of organic farming include advantages in sustainability, openness, self-sufficiency, autonomy/independence, health, food security, and food safety.

Organic agricultural methods are internationally regulated and legally enforced by many nations, based in large part on the standards set by the International Federation of Organic Agriculture Movements (IFOAM), an international umbrella organization for organic farming organizations established in 1972. Organic agriculture can be defined as:

An integrated farming system that strives for sustainability, the enhancement of soil fertility and biological diversity whilst, with rare exceptions, prohibiting synthetic pesticides, antibiotics, synthetic fertilizers, genetically modified organisms, and growth hormones.

Since 1990 the market for organic food and other products has grown rapidly, reaching $63 billion worldwide in 2012. This demand has driven a similar increase in organically managed farmland that grew from 2001 to 2011 at a compounding rate of 8.9% per annum.

As of 2019, approximately 70,000,000 hectares (170,000,000 acres) worldwide were farmed organically, representing approximately 1.4 percent of total world farmland.

Methods

"Organic agriculture is a production system that sustains the health of soils, ecosystems and people. It relies on ecological processes, biodiversity and cycles adapted to local conditions, rather than the use of inputs with adverse effects. Organic agriculture combines tradition, innovation and science to benefit the shared environment and promote fair relationships and a good quality of life for all involved."

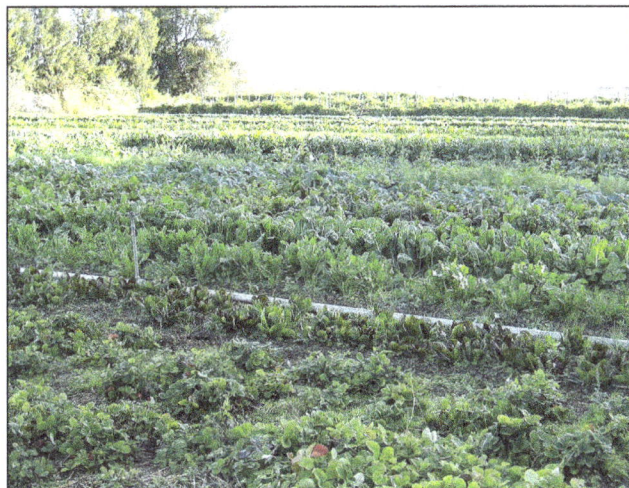

Organic cultivation of mixed vegetables in Capay.

Organic farming methods combine scientific knowledge of ecology and some modern technology with traditional farming practices based on naturally occurring biological processes. Organic farming methods are studied in the field of agroecology. While conventional agriculture uses synthetic pesticides and water-soluble synthetically purified fertilizers, organic farmers are restricted by regulations to using natural pesticides and fertilizers. An example of a natural pesticide is pyrethrin, which is found naturally in the Chrysanthemum flower. The principal methods of organic farming include crop rotation, green manures and compost, biological pest control, and mechanical cultivation. These measures use the natural environment to enhance agricultural productivity: legumes are planted to fix nitrogen into the soil, natural insect predators are encouraged, crops are rotated to confuse pests and renew soil, and natural materials such as potassium bicarbonate and mulches are used to control disease and weeds. Genetically modified seeds and animals are excluded.

While organic is fundamentally different from conventional because of the use of carbon based fertilizers compared with highly soluble synthetic based fertilizers and biological pest control instead of synthetic pesticides, organic farming and large-scale conventional farming are not entirely mutually exclusive. Many of the methods developed for organic agriculture have been borrowed by more conventional agriculture. For example, Integrated Pest Management is a multifaceted strategy that uses various organic methods of pest control whenever possible, but in conventional farming could include synthetic pesticides only as a last resort.

Crop Diversity

Organic farming encourages Crop diversity. The science of agroecology has revealed the benefits of polyculture (multiple crops in the same space), which is often employed in organic farming. Planting a variety of vegetable crops supports a wider range of beneficial insects, soil microorganisms, and other factors that add up to overall farm health. Crop diversity helps environments thrive and protects species from going extinct.

Soil Management

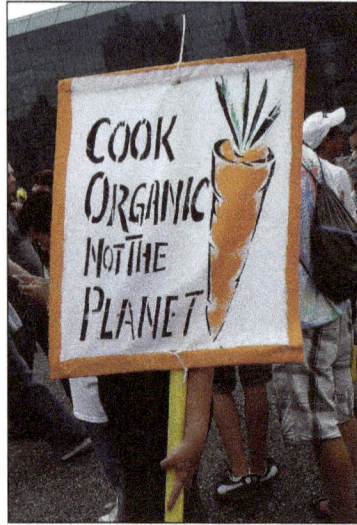

Placard advocating organic food rather than global warming.

Organic farming relies heavily on the natural breakdown of organic matter, using techniques like green manure and composting, to replace nutrients taken from the soil by previous crops. This biological process, driven by microorganisms such as mycorrhiza, allows the natural production of nutrients in the soil throughout the growing season, and has been referred to as *feeding the soil to feed the plant*. Organic farming uses a variety of methods to improve soil fertility, including crop rotation, cover cropping, reduced tillage, and application of compost. By reducing tillage, soil is not inverted and exposed to air; less carbon is lost to the atmosphere resulting in more soil organic carbon. This has an added benefit of carbon sequestration, which can reduce green house gases and help reverse climate change.

Plants need a large number of nutrients in various quantities to flourish. Supplying enough nitrogen and particularly synchronization so that plants get enough nitrogen at the when plants need it most, is a challenge for organic farmers. Crop rotation and green manure ("cover crops") help to provide nitrogen through legumes (more precisely, the *Fabaceae* family), which fix nitrogen from the atmosphere through symbiosis with rhizobial bacteria. Intercropping, which is sometimes used for insect and disease control, can also increase soil nutrients, but the competition between the legume and the crop can be problematic and wider spacing between crop rows is required. Crop residues can be ploughed back into the soil, and different plants leave different amounts of nitrogen, potentially aiding synchronization. Organic farmers also use animal manure, certain processed fertilizers such as seed meal and various mineral powders such as rock phosphate and green sand, a naturally occurring form of potash that provides potassium. Together these methods help to control erosion. In some cases pH may need to be amended. Natural pH amendments include lime and sulfur, but in the U.S. some compounds such as iron sulfate, aluminum sulfate, magnesium sulfate, and soluble boron products are allowed in organic farming.

Mixed farms with both livestock and crops can operate as ley farms, whereby the land gathers fertility through growing nitrogen-fixing forage grasses such as white clover or alfalfa and grows cash crops or cereals when fertility is established. Farms without livestock ("stockless") may find it more difficult to maintain soil fertility, and may rely more on external inputs such as imported manure as well as grain legumes and green manures, although grain legumes may fix limited nitrogen because they are harvested. Horticultural farms that grow fruits and vegetables in protected conditions often rely even more on external inputs.

Biological research into soil and soil organisms has proven beneficial to organic farming. Varieties of bacteria and fungi break down chemicals, plant matter and animal waste into productive soil nutrients. In turn, they produce benefits of healthier yields and more productive soil for future crops. Fields with less or no manure display significantly lower yields, due to decreased soil microbe community. Increased manure improves biological activity, providing a healthier, more arable soil system and higher yields.

Weed Management

Organic weed management promotes weed suppression, rather than weed elimination, by enhancing crop competition and phytotoxic effects on weeds. Organic farmers integrate cultural, biological, mechanical, physical and chemical tactics to manage weeds without synthetic herbicides.

Organic standards require rotation of annual crops, meaning that a single crop cannot be grown in the same location without a different, intervening crop. Organic crop rotations frequently include weed-suppressive cover crops and crops with dissimilar life cycles to discourage weeds associated with a particular crop. Research is ongoing to develop organic methods to promote the growth of natural microorganisms that suppress the growth or germination of common weeds.

Other cultural practices used to enhance crop competitiveness and reduce weed pressure include selection of competitive crop varieties, high-density planting, tight row spacing, and late planting into warm soil to encourage rapid crop germination.

Mechanical and physical weed control practices used on organic farms can be broadly grouped as:

- Tillage - Turning the soil between crops to incorporate crop residues and soil amendments; remove existing weed growth and prepare a seedbed for planting; turning soil after seeding to kill weeds, including cultivation of row crops;

- Mowing and cutting - Removing top growth of weeds;

- Flame weeding and thermal weeding - Using heat to kill weeds; and

- Mulching - Blocking weed emergence with organic materials, plastic films, or landscape fabric. Some critics, citing work published in 1997 by David Pimentel of Cornell University, which described an epidemic of soil erosion worldwide, have raised concerned that tillage contribute to the erosion epidemic. The FAO and other organizations have advocated a 'no-till' approach to both conventional and organic farming, and point out in particular that crop rotation techniques used in organic farming are excellent no-till approaches. A study published in 2005 by Pimentel and colleagues confirmed that 'Crop rotations and cover cropping (green manure) typical of organic agriculture reduce soil erosion, pest problems, and pesticide use.'

Some naturally sourced chemicals are allowed for herbicidal use. These include certain formulations of acetic acid (concentrated vinegar), corn gluten meal, and essential oils. A few selective bioherbicides based on fungal pathogens have also been developed. At this time, however, organic herbicides and bioherbicides play a minor role in the organic weed control toolbox.

Weeds can be controlled by grazing. For example, geese have been used successfully to weed a range of organic crops including cotton, strawberries, tobacco, and corn, reviving the practice of keeping cotton patch geese, common in the southern U.S. before the 1950s. Similarly, some rice farmers introduce ducks and fish to wet paddy fields to eat both weeds and insects.

Controlling other Organisms

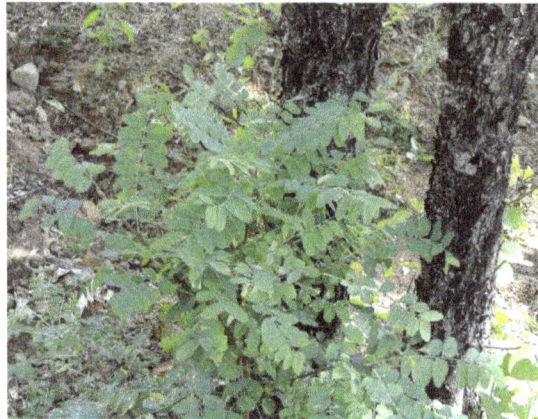

Chloroxylon is used for Pest Management in Organic Rice Cultivation.

Organisms aside from weeds that cause problems on organic farms include arthropods (e.g., insects, mites), nematodes, fungi and bacteria. Organic practices include, but are not limited to:

- Encouraging predatory beneficial insects to control pests by serving them nursery plants and/or an alternative habitat, usually in a form of a shelterbelt, hedgerow, or beetle bank;

- Encouraging beneficial microorganisms;

- Rotating crops to different locations from year to year to interrupt pest reproduction cycles;

- Planting companion crops and pest-repelling plants that discourage or divert pests;

- Using row covers to protect crops during pest migration periods;

- Using biologic pesticides and herbicides;

- Using stale seed beds to germinate and destroy weeds before planting;

- Using sanitation to remove pest habitat;

- Using insect traps to monitor and control insect populations; and

- Using physical barriers, such as row covers.

Examples of predatory beneficial insects include minute pirate bugs, big-eyed bugs, and to a lesser extent ladybugs (which tend to fly away), all of which eat a wide range of pests. Lacewings are also effective, but tend to fly away. Praying mantis tend to move more slowly and eat less heavily.

Parasitoid wasps tend to be effective for their selected prey, but like all small insects can be less effective outdoors because the wind controls their movement. Predatory mites are effective for controlling other mites.

Naturally derived insecticides allowed for use on organic farms use include *Bacillus thuringiensis* (a bacterial toxin), pyrethrum (a chrysanthemum extract), spinosad (a bacterial metabolite), neem (a tree extract) and rotenone (a legume root extract). Fewer than 10% of organic farmers use these pesticides regularly; one survey found that only 5.3% of vegetable growers in California use rotenone while 1.7% use pyrethrum. These pesticides are not always more safe or environmentally friendly than synthetic pesticides and can cause harm. The main criterion for organic pesticides is that they are naturally derived, and some naturally derived substances have been controversial. Controversial natural pesticides include rotenone, copper, nicotine sulfate, and pyrethrums Rotenone and pyrethrum are particularly controversial because they work by attacking the nervous system, like most conventional insecticides. Rotenone is extremely toxic to fish and can induce symptoms resembling Parkinson's disease in mammals. Although pyrethrum (natural pyrethrins) is more effective against insects when used with piperonyl butoxide (which retards degradation of the pyrethrins), organic standards generally do not permit use of the latter substance.

Naturally derived fungicides allowed for use on organic farms include the bacteria *Bacillus subtilis* and *Bacillus pumilus*; and the fungus *Trichoderma harzianum*. These are mainly effective for diseases affecting roots. Compost tea contains a mix of beneficial microbes, which may attack or out-compete certain plant pathogens, but variability among formulations and preparation methods may contribute to inconsistent results or even dangerous growth of toxic microbes in compost teas.

Some naturally derived pesticides are not allowed for use on organic farms. These include nicotine sulfate, arsenic, and strychnine.

Synthetic pesticides allowed for use on organic farms include insecticidal soaps and horticultural oils for insect management; and Bordeaux mixture, copper hydroxide and sodium bicarbonate for managing fungi. Copper sulfate and Bordeaux mixture (copper sulfate plus lime), approved for organic use in various jurisdictions, can be more environmentally problematic than some synthetic fungicides dissallowed in organic farming Similar concerns apply to copper hydroxide. Repeated application of copper sulfate or copper hydroxide as a fungicide may eventually result in copper accumulation to toxic levels in soil, and admonitions to avoid excessive accumulations of copper in soil appear in various organic standards and elsewhere. Environmental concerns for several kinds of biota arise at average rates of use of such substances for some crops. In the European Union, where replacement of copper-based fungicides in organic agriculture is a policy priority, research is seeking alternatives for organic production.

Livestock

Raising livestock and poultry, for meat, dairy and eggs, is another traditional farming activity that complements growing. Organic farms attempt to provide animals with natural living conditions and feed. Organic certification verifies that livestock are raised according to the USDA organic regulations throughout their lives. These regulations include the requirement that all animal feed must be certified organic.

For livestock, like these healthy cows, vaccines play an important part in animal
health since antibiotic therapy is prohibited in organic farming.

Organic livestock may be, and must be, treated with medicine when they are sick, but drugs cannot be used to promote growth, their feed must be organic, and they must be pastured.

Also, horses and cattle were once a basic farm feature that provided labor, for hauling and plowing, fertility, through recycling of manure, and fuel, in the form of food for farmers and other animals. While today, small growing operations often do not include livestock, domesticated animals are a desirable part of the organic farming equation, especially for true sustainability, the ability of a farm to function as a self-renewing unit.

Genetic Modification

A key characteristic of organic farming is the rejection of genetically engineered plants and animals. On 19 October 1998, participants at IFOAM's 12th Scientific Conference issued the Mar del Plata Declaration, where more than 600 delegates from over 60 countries voted unanimously to exclude the use of genetically modified organisms in food production and agriculture.

Although opposition to the use of any transgenic technologies in organic farming is strong, agricultural researchers Luis Herrera-Estrella and Ariel Alvarez-Morales continue to advocate integration of transgenic technologies into organic farming as the optimal means to sustainable agriculture, particularly in the developing world. Organic farmer Raoul Adamchak and geneticist Pamela Ronald write that many agricultural applications of biotechnology are consistent with organic principles and have significantly advanced sustainable agriculture.

Although GMOs are excluded from organic farming, there is concern that the pollen from genetically modified crops is increasingly penetrating organic and heirloom seed stocks, making it difficult, if not impossible, to keep these genomes from entering the organic food supply. Differing regulations among countries limits the availability of GMOs to certain countries, as described in the article on regulation of the release of genetic modified organisms.

Tools

Organic farmers use a number of traditional farm tools to do farming. Due to the goals of sustainability in organic farming, organic farmers try to minimize their reliance on fossil fuels. In the developing world on small organic farms tools are normally constrained to hand tools and diesel powered water pumps.

Standards

Standards regulate production methods and in some cases final output for organic agriculture. Standards may be voluntary or legislated. As early as the 1970s private associations certified organic producers. In the 1980s, governments began to produce organic production guidelines. In the 1990s, a trend toward legislated standards began, most notably with the 1991 EU-Eco-regulation developed for European Union, which set standards for 12 countries, and a 1993 UK program. The EU's program was followed by a Japanese program in 2001, and in 2002 the U.S. created the National Organic Program (NOP). As of 2007 over 60 countries regulate organic farming. In 2005 IFOAM created the Principles of Organic Agriculture, an international guideline for certification criteria. Typically the agencies accredit certification groups rather than individual farms.

Composting

Using manure as a fertilizer risks contaminating food with animal gut bacteria, including pathogenic strains of E. coli that have caused fatal poisoning from eating organic food. To combat this risk, USDA organic standards require that manure must be sterilized through high temperature thermophilic composting. If raw animal manure is used, 120 days must pass before the crop is harvested if the final product comes into direct contact with the soil. For products that don't directly contact soil, 90 days must pass prior to harvest.

Economics

The economics of organic farming, a subfield of agricultural economics, encompasses the entire process and effects of organic farming in terms of human society, including social costs, opportunity costs, unintended consequences, information asymmetries, and economies of scale. Although the scope of economics is broad, agricultural economics tends to focus on maximizing yields and efficiency at the farm level. Economics takes an anthropocentric approach to the value of the natural world: biodiversity, for example, is considered beneficial only to the extent that it is valued by people and increases profits. Some entities such as the European Union subsidize organic farming, in large part because these countries want to account for the externalities of reduced water use, reduced water contamination, reduced soil erosion, reduced carbon emissions, increased biodiversity, and assorted other benefits that result from organic farming.

Traditional organic farming is labor and knowledge-intensive whereas conventional farming is capital-intensive, requiring more energy and manufactured inputs. Organic farmers in California have cited marketing as their greatest obstacle.

Geographic Producer Distribution

The markets for organic products are strongest in North America and Europe, which as of 2001 are estimated to have $6 and $8 billion respectively of the $20 billion global market. As of 2007 Australasia has 39% of the total organic farmland, including Australia's 1,180,000 hectares (2,900,000 acres) but 97 percent of this land is sprawling rangeland. US sales are 20x as much. Europe farms 23 percent of global organic farmland (6,900,000 ha (17,000,000 acres)), followed by Latin America with 19 percent (5.8 million hectares - 14.3 million acres). Asia has 9.5 percent while North America has 7.2 percent. Africa has 3 percent.

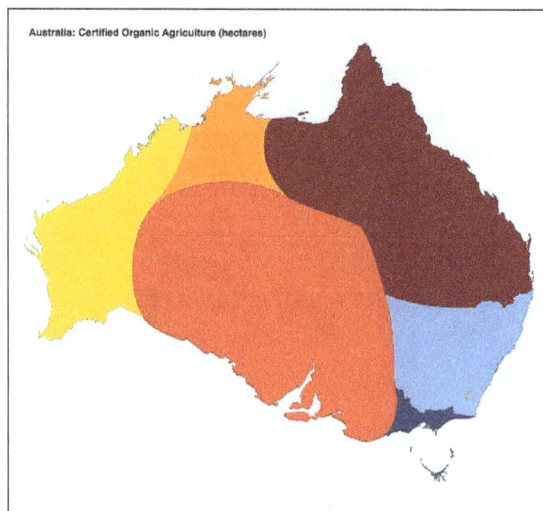

A density-equalising map of organic agriculture in Australia based on certified organic hectares. Australia accounts for more than half of the world's certified organic hectares.

Besides Australia, the countries with the most organic farmland are Argentina (3.1 million hectares - 7.7 million acres), China (2.3 million hectares - 5.7 million acres), and the United States (1.6 million hectares - 4 million acres). Much of Argentina's organic farmland is pasture, like that of Australia. Spain, Germany, Brazil (the world's largest agricultural exporter), Uruguay, and the England follow the United States in the amount of organic land.

In the European Union (EU25) 3.9% of the total utilized agricultural area was used for organic production in 2005. The countries with the highest proportion of organic land were Austria (11%) and Italy (8.4%), followed by the Czech Republic and Greece (both 7.2%). The lowest figures were shown for Malta (0.2%), Poland (0.6%) and Ireland (0.8%). In 2009, the proportion of organic land in the EU grew to 4.7%. The countries with highest share of agricultural land were Liechtenstein (26.9%), Austria (18.5%) and Sweden (12.6%). 16% of all farmers in Austria produced organically in 2010. By the same year the proportion of organic land increased to 20%.: In 2005 168,000 ha (415,000 ac) of land in Poland was under organic management. In 2012, 288,261 hectares (712,308 acres) were under organic production, and there were about 15,500 organic farmers; retail sales of organic products were EUR 80 million in 2011. As of 2012 organic exports were part of the government's economic development strategy.

After the collapse of the Soviet Union in 1991, agricultural inputs that had previously been purchased from Eastern bloc countries were no longer available in Cuba, and many Cuban farms converted to organic methods out of necessity. Consequently, organic agriculture is a mainstream practice in Cuba, while it remains an alternative practice in most other countries. Cuba's organic strategy includes development of genetically modified crops; specifically corn that is resistant to the palomilla moth.

Growth

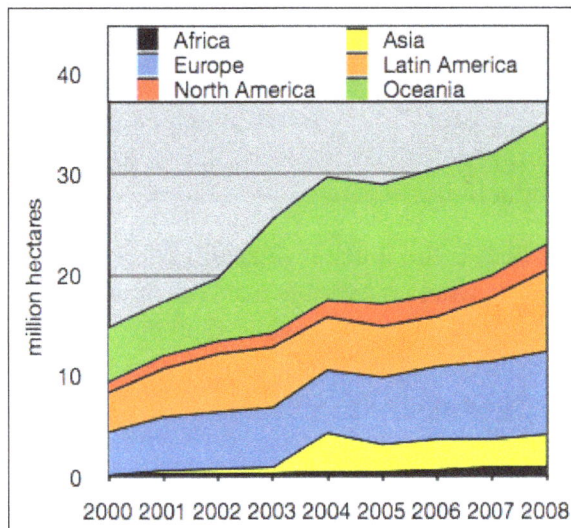

Organic farmland by world region.

In 2001, the global market value of certified organic products was estimated at USD $20 billion. By 2002, this was USD $23 billion and by 2015 more than USD $43 billion. By 2014, retail sales of organic products reached USD $80 billion worldwide. North America and Europe accounted for more than 90% of all organic product sales. In 2018 Australia accounted for 54% of the world's certified organic land with the country recording more than 35,000,000 verified organic hectares.

Organic agricultural land increased almost fourfold in 15 years, from 11 million hectares in 1999 to 43.7 million hectares in 2014. Between 2013 and 2014, organic agricultural land grew by 500,000 hectares worldwide, increasing in every region except Latin America. During this time period, Europe's organic farmland increased 260,000 hectares to 11.6 million total (+2.3%), Asia's increased 159,000 hectares to 3.6 million total (+4.7%), Africa's increased 54,000 hectares to 1.3 million total (+4.5%), and North America's increased 35,000 hectares to 3.1 million total (+1.1%). As of 2014, the country with the most organic land was Australia (17.2 million hectares), followed by Argentina (3.1 million hectares), and the United States (2.2 million hectares). Australia's organic land area has increased at a rate of 16.5% per annum for the past eighteen years.

In 2013, the number of organic producers grew by almost 270,000, or more than 13%. By 2014, there were a reported 2.3 million organic producers in the world. Most of the total global increase took place in the Philippines, Peru, China, and Thailand. Overall, the majority of all organic producers are in India (650,000 in 2013), Uganda (190,552 in 2014), Mexico (169,703 in 2013) and the Philippines (165,974 in 2014).

Productivity

Studies comparing yields have had mixed results. These differences among findings can often be attributed to variations between study designs including differences in the crops studied and the methodology by which results were gathered.

A 2012 meta-analysis found that productivity is typically lower for organic farming than conventional farming, but that the size of the difference depends on context and in some cases may be very small. While organic yields can be lower than conventional yields, another meta-analysis published in Sustainable Agriculture Research in 2015, concluded that certain organic on-farm practices could help narrow this gap. Timely weed management and the application of manure in conjunction with legume forages/cover crops were shown to have positive results in increasing organic corn and soybean productivity.

Another meta-analysis published in the journal Agricultural Systems in 2011 analyzed 362 datasets and found that organic yields were on average 80% of conventional yields. The author's found that there are relative differences in this yield gap based on crop type with crops like soybeans and rice scoring higher than the 80% average and crops like wheat and potato scoring lower. Across global regions, Asia and Central Europe were found to have relatively higher yields and Northern Europe relatively lower than the average.

A 2007 study compiling research from 293 different comparisons into a single study to assess the overall efficiency of the two agricultural systems has concluded that "organic methods could produce enough food on a global per capita basis to sustain the current human population, and potentially an even larger population, without increasing the agricultural land base." The researchers also found that while in developed countries, organic systems on average produce 92% of the yield produced by conventional agriculture, organic systems produce 80% more than conventional farms in developing countries, because the materials needed for organic farming are more accessible than synthetic farming materials to farmers in some poor countries. This study's methodology and results were contested by D. J. Connor of The University of Melbourne, in a short communication published in Field Crops Research. Connor writes that errors in Badgley et al. result in "major overestimation of the productivity of OA".

Long Term Studies

A study published in 2005 compared conventional cropping, organic animal-based cropping, and organic legume-based cropping on a test farm at the Rodale Institute over 22 years. The study found that "the crop yields for corn and soybeans were similar in the organic animal, organic legume, and conventional farming systems". It also found that "significantly less fossil energy was expended to produce corn in the Rodale Institute's organic animal and organic legume systems than in the conventional production system. There was little difference in energy input between the different treatments for producing soybeans. In the organic systems, synthetic fertilizers and pesticides were generally not used". As of 2013 the Rodale study was ongoing and a thirty-year anniversary report was published by Rodale in 2012.

A long-term field study comparing organic/conventional agriculture carried out over 21 years in Switzerland concluded that "Crop yields of the organic systems averaged over 21 experimental years at 80% of the conventional ones. The fertilizer input, however, was 34 – 51% lower, indicating an efficient production. The organic farming systems used 20 – 56% less energy to produce a crop unit and per land area this difference was 36 – 53%. In spite of the considerably lower pesticide input the quality of organic products was hardly discernible from conventional analytically and even came off better in food preference trials and picture creating methods".

Profitability

In the United States, organic farming has been shown to be 2.7 to 3.8 times more profitable for the farmer than conventional farming when prevailing price premiums are taken into account. Globally, organic farming is between 22 and 35 percent more profitable for farmers than conventional methods, according to a 2015 meta-analysis of studies conducted across five continents.

The profitability of organic agriculture can be attributed to a number of factors. First, organic farmers do not rely on synthetic fertilizer and pesticide inputs, which can be costly. In addition, organic foods currently enjoy a price premium over conventionally produced foods, meaning that organic farmers can often get more for their yield.

The price premium for organic food is an important factor in the economic viability of organic farming. In 2013 there was a 100% price premium on organic vegetables and a 57% price premium for organic fruits. These percentages are based on wholesale fruit and vegetable prices, available through the United States Department of Agriculture's Economic Research Service. Price premiums exist not only for organic versus nonorganic crops, but may also vary depending on the venue where the product is sold: farmers' markets, grocery stores, or wholesale to restaurants. For many producers, direct sales at farmers' markets are most profitable because the farmer receives the entire markup, however this is also the most time and labor-intensive approach.

There have been signs of organic price premiums narrowing in recent years, which lowers the economic incentive for farmers to convert to or maintain organic production methods. Data from 22 years of experiments at the Rodale Institute found that, based on the current yields and production costs associated with organic farming in the United States, a price premium of only 10% is required to achieve parity with conventional farming. A separate study found that on a global scale, price premiums of only 5-7% percent were needed to break even with conventional methods. Without the price premium, profitability for farmers is mixed.

For markets and supermarkets organic food is profitable as well, and is generally sold at significantly higher prices than non-organic food.

Energy Efficiency

In the most recent assessments of the energy efficiency of organic versus conventional agriculture, results have been mixed regarding which form is more carbon efficient. Organic farm systems have more often than not been found to be more energy efficient, however, this is not always the case. More than anything, results tend to depend upon crop type and farm size.

A comprehensive comparison of energy efficiency in grain production, produce yield, and animal husbandry concluded that organic farming had a higher yield per unit of energy over the vast majority of the crops and livestock systems. For example, two studies - both comparing organically- versus conventionally-farmed apples - declare contradicting results, one saying organic farming is more energy efficient, the other saying conventionally is more efficient.

It has generally been found that the labor input per unit of yield was higher for organic systems compared with conventional production.

Sales and Marketing

Most sales are concentrated in developed nations. In 2008, 69% of Americans claimed to occasionally buy organic products, down from 73% in 2005. One theory for this change was that consumers were substituting "local" produce for "organic" produce.

Distributors

The USDA requires that distributors, manufacturers, and processors of organic products be certified by an accredited state or private agency. In 2007, there were 3,225 certified organic handlers, up from 2,790 in 2004.

Organic handlers are often small firms; 48% reported sales below $1 million annually, and 22% between $1 and $5 million per year. Smaller handlers are more likely to sell to independent natural grocery stores and natural product chains whereas large distributors more often market to natural product chains and conventional supermarkets, with a small group marketing to independent natural product stores. Some handlers work with conventional farmers to convert their land to organic with the knowledge that the farmer will have a secure sales outlet. This lowers the risk for the handler as well as the farmer. In 2004, 31% of handlers provided technical support on organic standards or production to their suppliers and 34% encouraged their suppliers to transition to organic. Smaller farms often join together in cooperatives to market their goods more effectively.

93% of organic sales are through conventional and natural food supermarkets and chains, while the remaining 7% of U.S. organic food sales occur through farmers' markets, foodservices, and other marketing channels.

Direct-to-consumer Sales

In the 2012 Census, direct-to-consumer sales equaled $1.3 billion, up from $812 million in 2002, an increase of 60 percent. The number of farms that utilize direct-to-consumer sales was 144,530 in 2012 in comparison to 116,733 in 2002. Direct-to-consumer sales include farmers' markets, community supported agriculture (CSA), on-farm stores, and roadside farm stands. Some organic farms also sell products direct to retailer, direct to restaurant and direct to institution. According to the 2008 Organic Production Survey, approximately 7% of organic farm sales were direct-to-consumers, 10% went direct to retailers, and approximately 83% went into wholesale markets. In comparison, only 0.4% of the value of convention agricultural commodities were direct-to-consumers.

While not all products sold at farmer's markets are certified organic, this direct-to-consumer avenue has become increasingly popular in local food distribution and has grown substantially since 1994. In 2014, there were 8,284 farmer's markets in comparison to 3,706 in 2004 and 1,755 in 1994, most of which are found in populated areas such as the Northeast, Midwest, and West Coast.

Labor and Employment

Organic production is more labor-intensive than conventional production. On the one hand, this increased labor cost is one factor that makes organic food more expensive. On the other hand, the increased need for labor may be seen as an "employment dividend" of organic farming, providing more jobs per unit area than conventional systems. The 2011 UNEP Green Economy Report suggests that "an increase in investment in green agriculture is projected to lead to growth in employment of about 60 per cent compared with current levels" and that "green agriculture investments could create 47 million additional jobs compared with BAU2 over the next 40 years." The United Nations Environment Programme (UNEP) also argues that "by greening agriculture and food distribution, more calories per person per day, more jobs and business opportunities especially in rural areas, and market-access opportunities, especially for developing countries, will be available."

World's Food Security

In 2007 the United Nations Food and Agriculture Organization (FAO) said that organic agriculture often leads to higher prices and hence a better income for farmers, so it should be promoted. However, FAO stressed that by organic farming one could not feed the current mankind, even less the bigger future population. Both data and models showed then that organic farming was far from sufficient. Therefore, chemical fertilizers were needed to avoid hunger. Other analysis by many agribusiness executives, agricultural and environmental scientists, and international agriculture experts revealed the opinion that organic farming would not only increase the world's food supply, but might be the only way to eradicate hunger.

FAO stressed that fertilizers and other chemical inputs can much increase the production, particularly in Africa where fertilizers are currently used 90% less than in Asia. For example, in Malawi the yield has been boosted using seeds and fertilizers. FAO also calls for using biotechnology, as it can help smallholder farmers to improve their income and food security.

Also NEPAD, development organization of African governments, announced that feeding Africans and preventing malnutrition requires fertilizers and enhanced seeds.

According to a 2012 study in ScienceDigest, organic best management practices shows an average yield only 13% less than conventional. In the world's poorer nations where most of the world's hungry live, and where conventional agriculture's expensive inputs are not affordable by the majority of farmers, adopting organic management actually increases yields 93% on average, and could be an important part of increased food security.

Capacity Building in Developing Countries

Organic agriculture can contribute to ecological sustainability, especially in poorer countries. The application of organic principles enables employment of local resources (e.g., local seed varieties, manure, etc.) and therefore cost-effectiveness. Local and international markets for organic products show tremendous growth prospects and offer creative producers and exporters excellent opportunities to improve their income and living conditions.

Organic agriculture is knowledge intensive. Globally, capacity building efforts are underway, including localized training material, to limited effect. As of 2007, the International Federation of Organic Agriculture Movements hosted more than 170 free manuals and 75 training opportunities online.

In 2008 the United Nations Environmental Programme (UNEP) and the United Nations Conference on Trade and Development (UNCTAD) stated that "organic agriculture can be more conducive to food security in Africa than most conventional production systems, and that it is more likely to be sustainable in the long-term" and that "yields had more than doubled where organic, or near-organic practices had been used" and that soil fertility and drought resistance improved.

Millennium Development Goals

The value of organic agriculture (OA) in the achievement of the Millennium Development Goals (MDG), particularly in poverty reduction efforts in the face of climate change, is shown by its contribution to both income and non-income aspects of the MDGs. These benefits are expected to continue in the post-MDG era. A series of case studies conducted in selected areas in Asian countries by the Asian Development Bank Institute (ADBI) and published as a book compilation by ADB in Manila document these contributions to both income and non-income aspects of the MDGs. These include poverty alleviation by way of higher incomes, improved farmers' health owing to less chemical exposure, integration of sustainable principles into rural development policies, improvement of access to safe water and sanitation, and expansion of global partnership for development as small farmers are integrated in value chains.

A related ADBI study also sheds on the costs of OA programs and set them in the context of the costs of attaining the MDGs. The results show considerable variation across the case studies, suggesting that there is no clear structure to the costs of adopting OA. Costs depend on the efficiency of the OA adoption programs. The lowest cost programs were more than ten times less expensive than the highest cost ones. However, further analysis of the gains resulting from OA adoption reveals that the costs per person taken out of poverty was much lower than the estimates of the World Bank, based on income growth in general or based on the detailed costs of meeting some of the more quantifiable MDGs (e.g., education, health, and environment).

Externalities

Agriculture imposes negative externalities (uncompensated costs) upon society through public land and other public resource use, biodiversity loss, erosion, pesticides, nutrient runoff, subsidized water usage, subsidy payments and assorted other problems. Positive externalities include self-reliance, entrepreneurship, respect for nature, and air quality. Organic methods reduce some of these costs. In 2000 uncompensated costs for 1996 reached 2,343 million British pounds or £208 per ha (£84.20/ac). A study of practices in the US published in 2005 concluded that cropland costs the economy approximately 5 to 16 billion dollars ($30–96/ha – $12–39/ac), while livestock production costs 714 million dollars. Both studies recommended reducing externalities. The 2000 review included reported pesticide poisonings but did not include speculative chronic health effects of pesticides, and the 2004 review relied on a 1992 estimate of the total impact of pesticides.

It has been proposed that organic agriculture can reduce the level of some negative externalities from (conventional) agriculture. Whether the benefits are private or public depends upon the division of property rights.

Several surveys and studies have attempted to examine and compare conventional and organic systems of farming and have found that organic techniques, while not without harm, are less damaging than conventional ones because they reduce levels of biodiversity less than conventional systems do and use less energy and produce less waste when calculated per unit area.

Disadvantages

A 2003 to 2005 investigation by the Cranfield University for the Department for Environment, Food and Rural Affairs in the UK found that it is difficult to compare the Global warming potential, acidification and eutrophication emissions but "Organic production often results in increased burdens, from factors such as N leaching and N_2O emissions", even though primary energy use was less for most organic products. N_2O is always the largest global warming potential contributor except in tomatoes. However, "organic tomatoes always incur more burdens (except pesticide use)". Some emissions were lower "per area", but organic farming always required 65 to 200% more field area than non-organic farming. The numbers were highest for bread wheat (200+ % more) and potatoes (160% more).

Environmental Impact and Emissions

Researchers at Oxford University analyzed 71 peer-reviewed studies and observed that organic products are sometimes worse for the environment. Organic milk, cereals, and pork generated higher greenhouse gas emissions per product than conventional ones but organic beef and olives had lower emissions in most studies. Usually organic products required less energy, but more land. Per unit of product, organic produce generates higher nitrogen leaching, nitrous oxide emissions, ammonia emissions, eutrophication, and acidification potential than conventionally grown produce. Other differences were not significant. The researchers concluded that public debate should consider various manners of employing conventional or organic farming, and not merely debate conventional farming as opposed to organic farming. They also sought to find specific solutions to specific circumstances.

Proponents of organic farming have claimed that organic agriculture emphasizes closed nutrient cycles, biodiversity, and effective soil management providing the capacity to mitigate and even reverse the effects of climate change and that organic agriculture can decrease fossil fuel emissions. "The carbon sequestration efficiency of organic systems in temperate climates is almost double (575–700 kg carbon per ha per year – 510–625 lb/ac/an) that of conventional treatment of soils, mainly owing to the use of grass clovers for feed and of cover crops in organic rotations."

Critics of organic farming methods believe that the increased land needed to farm organic food could potentially destroy the rainforests and wipe out many ecosystems.

Nutrient Leaching

According to a 2012 meta-analysis of 71 studies, nitrogen leaching, nitrous oxide emissions, ammonia emissions, eutrophication potential and acidification potential were higher for organic products, although in one study "nitrate leaching was 4.4–5.6 times higher in conventional plots than organic plots". Excess nutrients in lakes, rivers, and groundwater can cause algal blooms, eutrophication, and subsequent dead zones. In addition, nitrates are harmful to aquatic organisms by themselves.

Land use

The Oxford meta-analysis of 71 studies found that organic farming requires 84% more land for an equivalent amount of harvest, mainly due to lack of nutrients but sometimes due to weeds, diseases or pests, lower yielding animals and land required for fertility building crops. While organic farming does not necessarily save land for wildlife habitats and forestry in all cases, the most modern breakthroughs in organic are addressing these issues with success.

Professor Wolfgang Branscheid says that organic animal production is not good for the environment, because organic chicken requires twice as much land as "conventional" chicken and organic pork a quarter more. According to a calculation by Hudson Institute, organic beef requires three times as much land. On the other hand, certain organic methods of animal husbandry have been shown to restore desertified, marginal, and/or otherwise unavailable land to agricultural productivity and wildlife. Or by getting both forage and cash crop production from the same fields simultaneously, reduce net land use.

In England organic farming yields 55% of normal yields. In other regions of the world, organic methods have started producing record yields.

Pesticides

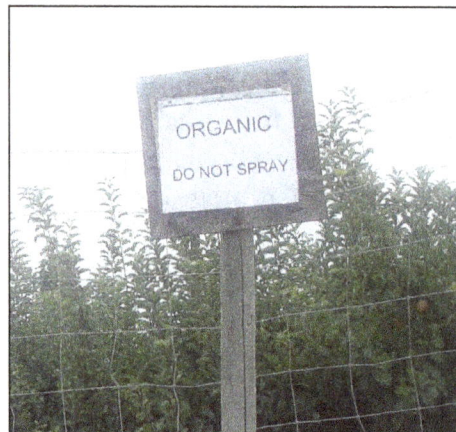

A sign outside of an organic apple orchard in Pateros, Washington reminding orchardists not to spray pesticides on these trees.

In organic farming synthetic pesticides are generally prohibited. A chemical is said to be synthetic if it does not already exist in the natural world. But the organic label goes further and usually prohibit compounds that exist in nature if they are produced by chemical synthesis. So the prohibition is also about the method of production and not only the nature of the compound.

A non-exhaustive list of organic approved pesticides with their median lethal doses:

- Copper(II) sulfate is used as a fungicide and is also used in conventional agriculture (LD50 300 mg/kg). Conventional agriculture has the option to use the less toxic Mancozeb (LD50 4,500 to 11,200 mg/kg).

- Boric acid is used as stomach poison that target insects (LD50: 2660 mg/kg).

- Pyrethrin comes from chemicals extracted from flowers of the genus Pyrethrum (LD50 of 370 mg/kg). Its potent toxicity is used to control insects.

- Lime sulfur (aka calcium polysulfide) and sulfur are considered to be allowed, synthetic materials (LD50: 820 mg/kg).

- Rotenone is a powerful insecticide that was used to control insects (LD50: 132 mg/kg). Despite the high toxicity of Rotenone to aquatic life and some links to Parkinson disease the compound is still allowed in organic farming as it is a naturally occurring compound.

- Bromomethane is a gas that is still used in the nurseries of Strawberry organic farming.

- Azadirachtin is a wide spectrum very potent insecticide. Almost non toxic to mammals (LD50 in rats is > 3,540 mg/kg) but affects beneficial insects.

Food Quality and Safety

While there may be some differences in the amounts of nutrients and anti-nutrients when organically produced food and conventionally produced food are compared, the variable nature of food production and handling makes it difficult to generalize results, and there is insufficient evidence to make claims that organic food is safer or healthier than conventional food. Claims that organic food tastes better are not supported by evidence.

Soil Conservation

Supporters claim that organically managed soil has a higher quality and higher water retention. This may help increase yields for organic farms in drought years. Organic farming can build up soil organic matter better than conventional no-till farming, which suggests long-term yield benefits from organic farming. An 18-year study of organic methods on nutrient-depleted soil concluded that conventional methods were superior for soil fertility and yield for nutrient-depleted soils in cold-temperate climates, arguing that much of the benefit from organic farming derives from imported materials that could not be regarded as self-sustaining.

In Dirt: the Erosion of Civilizations, geomorphologist David Montgomery outlines a coming crisis from soil erosion. Agriculture relies on roughly one meter of topsoil, and that is being depleted ten times faster than it is being replaced. No-till farming, which some claim depends upon pesticides, is one way to minimize erosion. However, a 2007 study by the USDA's Agricultural Research Service has found that manure applications in tilled organic farming are better at building up the soil than no-till.

Biodiversity

The conservation of natural resources and biodiversity is a core principle of organic production. Three broad management practices (prohibition/reduced use of chemical pesticides and inorganic fertilizers; sympathetic management of non-cropped habitats; and preservation of mixed farming) that are largely intrinsic (but not exclusive) to organic farming are particularly beneficial for farmland wildlife. Using practices that attract or introduce beneficial insects, provide habitat for birds and mammals, and provide conditions that increase soil biotic diversity serve to supply vital ecological services to organic production systems. Advantages to certified organic operations that implement these types of production practices include: Decreased dependence on outside fertility inputs; reduced pest management costs; more reliable sources of clean water; and better pollination.

Nearly all non-crop, naturally occurring species observed in comparative farm land practice studies show a preference for organic farming both by abundance and diversity. An average of 30%

more species inhabit organic farms. Birds, butterflies, soil microbes, beetles, earthworms, spiders, vegetation, and mammals are particularly affected. Lack of herbicides and pesticides improve biodiversity fitness and population density. Many weed species attract beneficial insects that improve soil qualities and forage on weed pests. Soil-bound organisms often benefit because of increased bacteria populations due to natural fertilizer such as manure, while experiencing reduced intake of herbicides and pesticides. Increased biodiversity, especially from beneficial soil microbes and mycorrhizae have been proposed as an explanation for the high yields experienced by some organic plots, especially in light of the differences seen in a 21-year comparison of organic and control fields.

Biodiversity from organic farming provides capital to humans. Species found in organic farms enhance sustainability by reducing human input (e.g., fertilizers, pesticides).

The USDA's Agricultural Marketing Service (AMS) announced the National Organic Program (NOP) final guidance on Natural Resources and Biodiversity Conservation for Certified Organic Operations. Given the broad scope of natural resources which includes soil, water, wetland, woodland and wildlife, the guidance provides examples of practices that support the underlying conservation principles and demonstrate compliance with USDA organic regulations 205.200. The final guidance provides organic certifiers and farms with examples of production practices that support conservation principles and comply with the USDA organic regulations, which require operations to maintain or improve natural resources. The final guidance also clarifies the role of certified operations (to submit an OSP to a certifier), certifiers (ensure that the OSP describes or lists practices that explain the operator's monitoring plan and practices to support natural resources and biodiversity conservation), and inspectors (onsite inspection) in the implementation and verification of these production practices.

A wide range of organisms benefit from organic farming, but it is unclear whether organic methods confer greater benefits than conventional integrated agri-environmental programs. Organic farming is often presented as a more biodiversity-friendly practice, but the generality of the beneficial effects of organic farming is debated as the effects appear often species- and context-dependent, and current research has highlighted the need to quantify the relative effects of local- and landscape-scale management on farmland biodiversity. There are four key issues when comparing the impacts on biodiversity of organic and conventional farming: It remains unclear whether a holistic whole-farm approach (i.e. organic) provides greater benefits to biodiversity than carefully targeted prescriptions applied to relatively small areas of cropped and/or non-cropped habitats within conventional agriculture (i.e. agri-environment schemes); Many comparative studies encounter methodological problems, limiting their ability to draw quantitative conclusions; Our knowledge of the impacts of organic farming in pastoral and upland agriculture is limited; There remains a pressing need for longitudinal, system-level studies in order to address these issues and to fill in the gaps in our knowledge of the impacts of organic farming, before a full appraisal of its potential role in biodiversity conservation in agroecosystems can be made.

Opposition to Labor Standards

Organic agriculture is often considered to be more socially just and economically sustainable for farmworkers than conventional agriculture. However, there is little social science research or consensus as to whether or not organic agriculture provides better working conditions than conventional agriculture. As many consumers equate organic and sustainable agriculture with small-scale, family-owned organizations it is widely interpreted that buying organic supports better conditions for farmworkers than buying with conventional producers. Organic agriculture is generally more labor-intensive due to its dependence on manual practices for fertilization and pest removal and relies heavily upon hired, non-family farmworkers rather than family members. Although illnesses from synthetic inputs pose less of a risk, hired workers still fall victim to debilitating musculo-skeletal disorders associated with agricultural work. The USDA certification requirements outline growing practices and ecological standards but do nothing to codify labor practices. Independent certification initiatives such as the Agricultural Justice Project, Domestic Fair Trade Working Group, and the Food Alliance have attempted to implement farmworker interests but because these initiatives require voluntary participation of organic farms, their standards cannot be wide-ly enforced. Despite the benefit to farmworkers of implementing labor standards, there is little support among the organic community for these social requirements. Many actors of the organic industry believe that enforcing labor standards would be unnecessary, unacceptable, or unviable due to the constraints of the market.

Types of Tillage

Tillage operations are broadly grouped into two types based on the time.

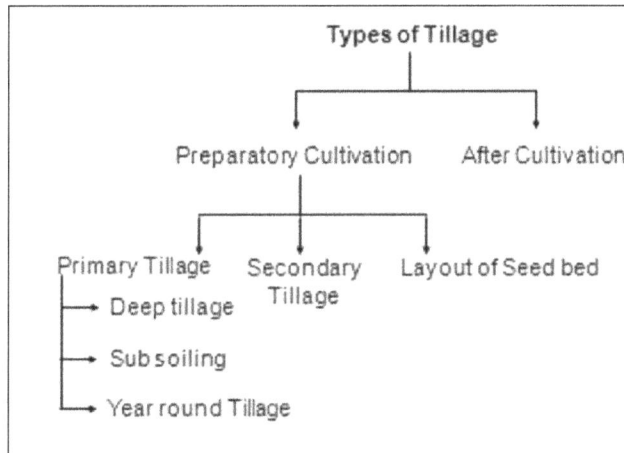

Types of Primary Tillage

Depending upon the purpose or necessity, different types of tillage are carried out. They are deep ploughing, subsoiling and year-round tillage.

Deep Tillage

Deep ploughing turns out large sized clods, which are baked by the hot sun when it is done in summer. These clods crumble due to alternate heating and cooling and due to occasional summer showers. This process of gradual disintegration of clods improves soil structure. The rhizomes and tubers of perennial weeds die due to exposure to hot sun. Summer deep ploughing kills pests due to exposure of pupae to hot sun.

A deep tillage of 25-30 cm depth is necessary for deep rooted crop like pigeonpea while moderate deep tillage of 15-20 cm is required for maize.

Deep tillage also improves soil moisture content. However the advantage of deep tillage in dry farming condition depends on rainfall pattern and crop. It is advisable to go for deep ploughing only for long duration, deep rooted crops. Depth of ploughing should be related to the amount of rainfall that it can wet.

Subsoiling

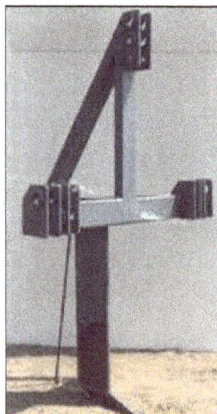

Hard pans may be present in the soil which restrict root growth of crops. These may be silt pans, iron or aluminium pans, clay pans or -man-made pans. Man-made pans are tillage pans induced by repeated tillage at the same depth. Root growth of crops is confined to top few centimetres of soil where deep penetration of roots is inhibited by hard pans.

For example, cotton roots grow to a depth of 2 m in deep alluvial soil without any pans. When hard pans are present, they grow only up to hard pan, say 15-20 cm. Similarly, vertical root growth of sugarcane is restricted due to hard pans and it is not compensated by horizontal spread. Subsoiling is breaking the hard pan without inversion and with less disturbance of top soil. A narrow cut is made in the top soil while share of the subsoiler shatters hard pans. Chisel ploughs are also used to break hard pans present even at 60-70 cm. The effect of subsoiling does not last long. To avoid closing of subsoil furrow, vertical mulching is adopted.

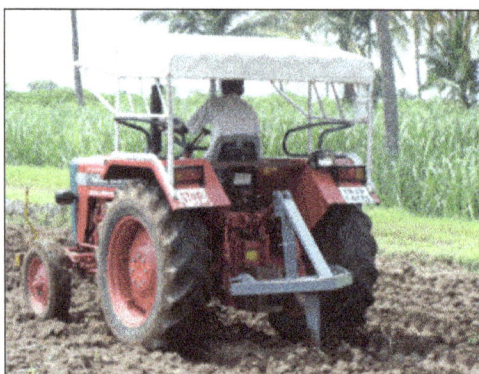

Year-round Tillage

Tillage operations carried out throughout the year are known as year-round tillage. In dry farming regions, field preparation is initiated with the help of summer showers. Repeated tillage operations are carried out until sowing of the crop. Even after harvest of the crop, the field is repeatedly ploughed or harrowed to avoid weed growth in the off season.

Secondary Tillage

Lighter or finer operations performed on the soil after primary tillage are known as secondary tillage. After ploughing, the fields are left with large clods with some weeds and stubbles partially uprooted.

Harrowing is done to a shallow depth to crush the clods and to uproot the remaining weeds and stubbles. Disc harrows, cultivators, blade harrows etc., are used for this purpose.

Planking is done to crush the hard clods to smoothen the soil surface and to compact the soil lightly. Thus the field is made ready for sowing after ploughing by harrowing and planking. Generally sowing operations are also included in secondary tillage.

Layout of Seedbed and Sowing

After the seedbed preparation, the field is laid out properly for irrigation and sowing or planting seedlings. These operations are crop specific. For most of the crops like wheat, soybean, pearl millet, groundnut, castor etc., flat levelled seedbed is prepared. After the secondary tillage, these crops are sown without any land treatments. However, growing crops during rainy season in deep black soils is a problem due to ill-drained conditions and as tillage is not possible during the rainy season. Broadbed and furrows (BBF) are, therefore, formed before the onset of monsoon and dry sowing is resorted to.

For some crops like maize, vegetables etc., the field has to be laid out into ridges and furrows. Sugarcane is planted in the furrows or trenches. Crops like tobacco, tomato, chillies are planted with equal inter and intra-row spacing so as to facilitate two-way intercultivation. After field preparation, a marker is run in both the directions. The seedlings are transplanted at the intercepts.

Layout of Seedbed

After Cultivation

The tillage operations that are carried out in the standing crop are called after tillage. It includes drilling or side dressing of fertilisers, earthing up and intercultivation.

Earthing up is an operation carried out with country plough or ridge plough so as to form ridges at the base of the crop. It is done either to provide extra support against lodging as in sugarcane or to provide more soil volume for better growth of tubers as in potato or to facilitate irrigation as in vegetables.

Intercultivation is working blade harrows, rotary hoes etc., in between the crop rows so as to control weeds. Intercultivation may also serve as moisture conservation measure by closing deep cracks in black soils.

TERRACE FARMING

In agriculture, a terrace is a piece of sloped plane that has been cut into a series of successively receding flat surfaces or platforms, which resemble steps, for the purposes of more effective farming. This type of landscaping is therefore called terracing. Graduated terrace steps are commonly used to farm on hilly or mountainous terrain. Terraced fields decrease both erosion and surface runoff, and may be used to support growing crops that require irrigation, such as rice. The Rice Terraces of the Philippine Cordilleras have been designated as a UNESCO World Heritage Site because of the significance of this technique.

Terraced paddy fields are used widely in rice, wheat and barley farming in east, south, and southeast Asia, as well as the Mediterranean, Africa, and South America. Drier-climate terrace farming is common throughout the Mediterranean Basin, where they are used for vineyards, olive trees, cork oak, etc.

In the South American Andes, farmers have used terraces, known as *andenes,* for over a thousand years to farm potatoes, maize, and other native crops. Terraced farming was developed by the Wari culture and other peoples of the south-central Andes before 1000 AD, centuries before they were used by the Inca, who adopted them. The terraces were built to make the most efficient use of shallow soil and to enable irrigation of crops by allowing runoff to occur through the outlet.

The Inca built on these, developing a system of canals, aqueducts, and puquios to direct water through dry land and increase fertility levels and growth. These terraced farms are found wherever mountain villages have existed in the Andes. They provided the food necessary to support the populations of great Inca cities and religious centres such as Machu Picchu.

Terracing is also used for sloping terrain; the Hanging Gardens of Babylon may have been built on an artificial mountain with stepped terraces, such as those on a ziggurat. At the seaside Villa of the Papyri in Herculaneum, the villa gardens of Julius Caesar's father-in-law were designed in terraces to give pleasant and varied views of the Bay of Naples.

Terraced fields are common in islands with steep slopes. The Canary Islands present a complex system of terraces covering the landscape from the coastal irrigated plantations to the dry fields in the highlands. These terraces, which are named *cadenas* (chains), are built with stone walls of skillful design, which include attached stairs and channels.

Tetang Village terraced fields, Mustang District.

Terraced fields in Sa Pa, Vietnam.

AGRONOMIC PRACTICES FOR THE MANAGEMENT OF SEED

Seed, particularly the new seed, viz., the High Yielding Varieties (HYV) which brought green revolution in the South East Asian countries like India, Bangladesh and Pakistan. The new seed or the HYV demands sound tillage practices, since the yield is much higher than the traditional varieties it demands added cost in harvesting, transportation, storage.

The new seeds are expensive in the initial stages of its introduction but the return is good for those who adopt it first but with increase in demand the returns get reduced as a matter of competition. In striving for economic balance we always run into economic principles, the product and return from a resource depend upon the amount of other resources and practices with which it is combined.

The HYV has revolutionized the agronomic practices known as packages of practice that is, the HYV demands application of high dosages of fertilizers, irrigation, plant protection measures and mechanization of tillage practices for the proper timings of practices.

Varietal trial should be given to allocate them to different regions suitable to these varieties. Farmers and Farm managers should keep in touch with the Agricultural Universities and Agricultural institutions for further information about new varieties, agronomic practices, yields levels, and economics of their production.

Managers of farms should consider opportunity cost principles for making right decisions for the investment of limited funds. Treatment against diseases, like late blight, smut, rust, red rot of sugarcane.

If treatment is given where the diseases do not occur then it would be an economic loss. In case the manager fails to adopt prophylactic measures and diseases appear it will result in significant loss and would hurt the economy. With the proper sowing and planting practices the cost per unit of output shall be reduced as the yield per unit of land will be significantly higher.

Time of sowing and planting should be religiously adhered to as there is optimum condition of air, moisture and temperature at the right time which are the requirement of crops. So the right time could be managed with the help of calendar of operation.

On a mechanized farm sowing and planting machinery would be profitable to add in the inventory but its annual use should be kept in mind otherwise custom hiring will be helpful. This will ensure timely operation. The operator of a large sized farm and larger heads of livestock having competitive demand for the labour and farm machinery should consider the opportunity cost principle and also another principle viz., the added cost and added returns.

The rates of sowing—this varies from crop to crop. The scientists do recommend certain quantity per hectare which varies from variety to variety. Excessive rate causes diminishing rates of return. Fertilizer quantity and seed rates have interrelationship.

The farm manager should make judgment of the proportion of the two. The best seeding rate will depend on the prices of the fertilizer and the product (yield of crop). Again, the calculation of the most profitable rate must consider the costs of added fertilizer and added seed in relation to the added yield and its value. In this context the germination rate is also important. The tillage operation as well as is considered on the principles of added cost and added returns.

The rate of application of fertilizer depends on crop variety, moisture content of the soil, seeding rate, application of irrigation and conservation practices.

The economic principle applied how much fertilizer should be applied are: Cost and Return principles (diminishing return and marginal cost), for farm with limited capital position opportunity cost and substitution principles are used for the combination of fertilizer with organic manure.

The quantity of fertilizer applied depends on management practices, price outlook, capital position and also the amount of other elements. There is no standard rate of application. The economic implication is the rates of return and price of products and the cost of fertilizer. Thus, it is wise to invest when the return is the highest.

The economic evolution of the time and methods of application of fertilizer must consider the following possibilities:

1.   If yields are the same under the two systems, compare the cost of application and select the least cost method.

2.   If the two costs are the same select the time and method which gives the greatest yield.

Partial budgeting is resorted to taking into consideration the cost of fertilizer, its application and return from the crop.

RELATIONSHIP OF AGRONOMIC PRACTICES TO SOIL NITROGEN DYNAMICS

Nitrogen (N) is the most important plant mineral nutrient. It was first discovered in the late eighteenth century, and N's role in improving crop production was widely recognized by the mid-nineteenth century. Long before these discoveries, ancient farmers often unknowingly employed agronomic practices that resulted in managing soil N availability, thereby helping to ensure the human food supply and nutrition. There were two major sources of N in agroecosystems before synthetic N fertilizers—soil N and legume-based biological N fixation. Ancient farmers constructively developed tillage schemes and rotated nonlegume and legume crops to manage both N sources for millennia. Because the appearance of commercial synthetic N fertilizers in the early twentieth century brought significant changes to traditional agronomic practices, the history of agronomic practices from the perspective of managing soil and biologically fixed N dynamics would seem to be a fruitful review.

Plow tillage is a form of soil N management. Much of the soil N is in complex organic forms, such as decomposing plant and animal residues. Most plants can only take up inorganic N (NH_4+ and NO_3-) although the basic amino acids are absorbed by some plant species (e.g., Picea abies). Inorganic N and basic amino acids in soil are mainly derived from N mineralization processes. Tillage practices can promote mineralization because disturbance exposes naturally protected (i.e., aggregate-protected) soil organic matter (SOM) to microbes, enhancing microbial activity and N mineralization (Tisdall and Oades). Therefore, plow tillage was considered a great agricultural advance and, from the archeological evidence, has had a very long history. Foot plows also called "digging sticks" are shown in Egyptian tomb paintings. A wooden model of oxen and plow was found in an Egyptian tomb dating from 2000 BCE. In Asia, one of the oldest existing Chinese books titled "Lü Shi Chun Qiu" or "The Annals of Lu Buwei" demonstrated the details of when and how to till according to soil and weather conditions and served as an early example of a practical farming guide.

Rotation can also be a tool to manage soil N through legume bio-fixation of N, depending on the crop species. Monocropping, especially with nonlegumes and heavy-nutrient-using crops (e.g., tobacco and corn) can deplete soil N. Rotation practices, even simple fallow, help to restore soil N. This practice was evident in early Roman times. One of the Rome's greatest poets, Virgil, wrote in his poem Georgics "For the field is drained by flax-harvest and wheat-harvest, drained by the slumber-steeped poppy of Lethe, but yet rotation lightens the labour." This emphasizes that fallow was necessary to rotate with those crops requiring more nutrients. On the other hand, rotations that include a legume crop can bring biological N fixation into agricultural production systems. Although ancient farmers knew nothing of the biological N fixation process, and nothing about the importance of mineral N to plant growth, they intentionally included legume crops into crop sequences. This was evident in the Pliny the Elder's book on natural history that mentioned several legume successions as alternatives to conditions that forbade fallowing.

Synthetic fertilizer N application in agricultural production has a relatively short history compared to tillage and rotation practices because knowledge regarding N in plant nutrition and N synthesis techniques is recent. In 1836, Jean-Baptiste Boussingault investigated manure, crop rotation, and N sources and for the first time concluded that N was a major component of plants and that the nutritional value of fertilizer was proportional to its N content. However, ammonia could not be easily synthesized from constituent elements until 1908, when the Haber-Bosch process was developed. After that, synthetic fertilizer N started to play a greater role in agricultural production, helping to improve global food security.

Influence of Synthetic Fertilizer N on Traditional Agronomic Practices

The appearance of synthetic fertilizer N brought a huge increase in the global food supply. Erisman et al. estimated that around 50% of the world population's food requirements are currently met by using synthetic fertilizer N. However, synthetic fertilizer N fundamentally disturbed the soil N cycling balance in agro-ecosystems and brought significant changes to traditional agronomic practices. Our unpublished data show that synthetic fertilizer N can promote or prime soil N mineralization depending on the indigenous SOM level and the amount of synthetic N.

Synthetic fertilizer N played a role in developing modern no-tillage farming. Agriculture derives numerous benefits from no-tillage, including fuel and labor savings, increased soil C stocks and erosion resistance. But few people recognized the fertilizer N contribution to no-tillage until early Kentucky no-tillage × N fertility trials revealed its importance. No-tillage without N fertilizer significantly lowered yield compared to conventional tillage without N fertilizer. However, no-tillage with N fertilizer produced yields comparable to those of conventional tillage with fertilizer N. From this perspective, one can speculate that added fertilizer N compensated for reduced soil N mineralization in no-tillage. Other factors, including herbicide and equipment development, also made no-tillage farming feasible in Kentucky and the rest of the Southeast and mid-Atlantic states in the USA, beginning in the 1960s. At the time, the move away from tillage was viewed with much skepticism, but eventually no-tillage was accepted as a revolution in farming. By 2009, approximately 36% of U.S. cropland, planted to eight major crops, was in no-tillage soil management.

Although ancient farmers knew nothing of biological N fixation, legume crops had been an important cropping system component worldwide before synthetic N became available. However, crop rotation was discouraged during the Green Revolution, partially because pest control benefits from crop rotation could be replaced by chemical crop protectants. Also, the N credits from biological N fixation could be easily replaced by synthetic fertilizer N. However, soon after the height of the Green Revolution, many studies reported that no amount of chemical fertilizer or pesticide could fully compensate for crop rotation benefits. Rotation systems then came back into fashion. Currently, 80% of all corn, soybean, and wheat planted acres in the United States are in rotation.

Systematic understanding of Agronomic Practices and Soil N Dynamics

Agricultural history shows that managing N dynamics is one of the central reasons farmers developed and implemented specific agronomic practices. Furthermore, in the last few decades, new knowledge indicates how transient N can have negative impacts on global environments and human health. A systematic understanding of "How does soil and crop sequence management influence nitrogen dynamics?" will significantly influence agronomic practice development but also has global meaning for the quality of human life. The aim of optimal agricultural N management is to enhance net N mineralization at times when crops need N, to synchronize soil N mineralization with crop N uptake, and to minimize N loss. To systematically understand, three sequential steps need clarification:

- How do agronomic practices affect soil organic matter pools?

- How do soil organic matter pools contribute to soil N availability?

- How do agronomic practices influence crop N uptake capacity?

Soil organic and crop residue N pools provide the organic N for N mineralization. This microbial process, primarily heterotrophic, also requires soil organic C (SOC) as an energy source. Thus, to understand how soil and crop management affect mineralized soil N, it is critical to first evaluate whether and how tillage, rotation, and fertilizer N application affect SOC and N sequestration. Soil organic matter sequestration has been reported to be linked with soil aggregate formation. The dominant concept that explains SOC and N sequestration is based on the aggregation-SOM model. Zou et al. reported that using NT and/or rotation practices in burley tobacco production maintained desirable soil physical and chemical properties by macroaggregate stabilization, which led to conserving SOC and TSN stocks. The basic idea is that soil organic matter functions as a nucleus/binding agent for aggregate formation. Aggregates are important reservoirs of SOC and N that are protected from microbial access and less subject to physical, chemical, microbial, and enzymatic degradation.

Appropriate and precise estimation of soil N mineralization has been a challenge since the early 1900s. Temporal and spatial variability are large because this process is determined by internal soil factors (e.g., SOM level, labile C and N pools, soil microbial community) and external environment factors (e.g., temperature, precipitation and aeration). Agronomic management, such as plant species and N fertilizer application, may also affect N mineralization. With current technologies, it is impossible to predict N mineralization by taking these factors into consideration simultaneously. Instead of being a measure of available N supply, N mineralization estimates by current methods should be considered an index of N availability.

Isotopic tracers and incubation methods are the two main approaches used to estimate N mineralization. The isotopic tracer method can measure gross N mineralization, but isotope methods are expensive and can also have methodological problems with mineralization rate estimates and other assumption violations. Although incubation methods can only measure net soil N mineralization (net soil N mineralization = gross N mineralization–N immobilization), incubation can fairly estimate the available N pool, which has a practical value for efficient N management in agroecosystems. Therefore, long-term biological mineralization has been considered the most suitable soil N availability index and is often used to validate other indices derived from more rapid chemical or biological assays. There are, however, many variations to incubation methods, including environment, sample pretreatment, and incubation time, and each variation has advantages and disadvantages. To use incubation to meet research objectives, assumptions, benefits, and liabilities of each variation should be considered.

An experimentally derived N availability index might not necessarily reflect total crop N uptake. Besides the amount of available soil N, crop N accumulation also depends on N uptake capacity. Crop N uptake capacity might be determined by either/both genetic and environmental controls. Genetics can control crop growth rate and biomass accumulation, which would result in different N demands at different growth stages. Crop species have different root architectures, mostly controlled by genetics. However, roots, the dominant nutrient uptake organ directly exposed to the soil, interact with a wide array of soil physical, chemical, and biological factors that vary in time and space. To understand the impact of agronomic management practices on crop N uptake or yield, both soil N availability and root architecture must be considered.

The effect of agronomic practices on crop N uptake or yield is reviewed in three sequential steps. First, the mechanism and effect of agronomic practices on SOC and STN sequestration are described. Second, the pros and cons of long-term incubation methodologies for estimating N mineralization are described. Finally, the potential effects of soil and crop management.

Mechanisms and Effect of Agronomic Practices on soil C and N Sequestration

The link between SOC and total soil N (STN) decomposition and stabilization and soil aggregate dynamics has been developed, recognized, and intensively studied since the 1900s. Soil organic C and N dynamics are important to agricultural production because these affect soil nutrient cycling and plant productivity. The C and N dynamics are also important to the environment because they can affect greenhouse gas emissions and water quality. These processes happen in a heterogeneous soil matrix and have multiple interactions with soil biota. The task of elucidation is complex. Aggregate-SOM models can explain some of these complexities. Aggregates not only physically protect SOC and SON, but also influence soil microbe community structure, limit oxygen diffusion, regulate water flow, determine nutrient adsorption and desorption, and reduce surface runoff and erosion. All these processes have fundamental effects on soil C and N sequestration and stabilization.

More current studies to understand the impact of agronomic practices on soil C and N sequestration have been based on the aggregate hierarchy concept proposed and developed by Tisdall and Oades. To apply the theoretical aggregate-SOM models, the first consideration is the physical separation of soil into different aggregate size classes. Two main methods to separate soil aggregates are widely used by researchers: dry and wet sieving. The disruption of aggregates is mainly due to slaking and microcracking when the soil is initially dry. Dry sieving of air-dried samples is used to characterize the aggregate size distribution with minimum destruction. Wet sieving is used to simulate microcracking and slaking. Water-stable aggregate stability from wet sieving procedures was reported to be closely correlated with SOM stabilization because SOM can act as a transient binding agent and has served as an effective early indicator of soil C change in numerous studies. The wet sieving procedure has been frequently used to evaluate the agronomic practice effects on both SOM sequestration and soil structural stability. In the wet sieving procedure, sample pretreatment is important. The rewetting pretreatments for soils can cause different results when comparing soils and management history treatments. Cambardella and Elliott showed that capillary-wetted soils retained more macroaggregates (>250 μm) than slaked soils. Bissonnais demonstrated that the different aggregate breakdown methods and frequency of crusting soil samples can dramatically affect soil aggregate stability within the same soil management system. Adopting minimum breakdown aggregates in the sieving procedure keeps comparisons between treatments relative to the natural field conditions.

The effect of agronomic practices (including tillage, rotation, and fertilizer N application) on SOC and STN, according to aggregate-SOM models, has been studied intensively in grass and grain crop production systems, but not in leaf harvest crop production systems. In these studies, no-tillage increased or maintained SOC and STN compared to conventional tillage. With aggregate separation, conventional tillage can increase large aggregate turnover rate, diminishing the macroaggregate proportion and SOC and STN concentrations. In contrast, no-tillage increases macroaggregates and SOC and STN accumulation.

Most studies show that rotation increases SOC and STN sequestration, compared to monocropping.

Crops in rotation schemes have different impacts on SOM stabilization, depending on the quantity and quality of crop residues. Wright and Hons found that crop residue production was similar among wheat, sorghum, and soybean fields, but the wheat field had significantly higher SOC and STN in surface soil than the other two fields, which indicates that the higher C:N ratio in wheat residue can play a role in SOM stabilization. Kong et al. reported that the quantity of crop residue/carbon production had a linear relationship with SOC sequestration in sustainable cropping systems. Therefore, when evaluating crop rotation schemes on SOM sequestration, examining crop residue quantity and quality is important.

Studies on the effect of fertilizer N application on SOM sequestration have produced the most controversial results. Some studies report that fertilizer N application increases SOM because higher fertilizer N input causes more crop residue to be returned to soil. Mulvaney et al. reported that fertilizer N application decreased soil N in the long-term Morrow plot study and argued that synthetic N application enhanced soil microbial decomposition due to the decreasing C:N ratio. Others have found no effect of N fertilizer application on SOM sequestration.

Methodologies of Soil N Mineralization Measurement

There are many different methods available for long-term aerobic incubation, in laboratory and field, depending on soil sample pretreatment and other incubation conditions.

Laboratory Incubation Methods

Most aerobic laboratory incubation methods have common features, including maintenance of optimal soil water status (typically with 60% water-filled pore space), constant temperature (commonly 25, 30, or 35 °C), and periodic sampling to estimate N mineralization rates. Although there have been several standardized protocols, there is significant variation in aerobic incubation details.

Leaching versus Non-leaching Processes

In early studies with long-term N mineralization incubation, samples were usually incubated continuously in a container without periodic leaching of the accumulated inorganic N. The merit of this method was convenience, but there could be cumulative inhibitory effects, such as pH decline, on mineralization during the incubation. Thus, nonleaching approaches were not recommended for long incubation periods. Stanford and Hanway proposed a periodic leaching approach during incubation. Briefly, 0.01 M $CaCl_2$ was used to leach mineralized N from the sample at the end of each incubation period. The merit of leaching would be avoidance of accumulation of unspecified toxins. While being a time-consuming and apparatus-requiring process, there was also an additional technical concern with potential leaching of soluble organic N during the incubation.

Excluded Crop Residue versus Included Crop Residue

Crop residue can contribute to the soil inorganic N pool by N mineralization or immobilization, depending on the residue C:N ratio. Most laboratory incubation methods exclude such contributions by discarding visible pieces of residue in the pretreatment sieving process. Some laboratory methods cut entrained residues into pieces that are mixed with soil for incubation. Certainly, discarding large portions of residue might influence estimates of the N credit from the previous crop because soil fertility guidelines usually recommend a different fertilizer N rate for the current crop that depends on the previous crop.

Field-moist Soil Sample versus Dried and Ground Soil Sample

Using dried and/or ground soil is convenient for a large amount of soil samples that require time to process or for cooperative projects where soil samples come from multiple locations at different times. However, several days are needed to rewet soil for preincubation, which also causes an N mineralization flush during the first weeks of incubation. Numerous studies report that sample sieving and drying-rewetting causes rapid microbial death and enhances microbial respiration and activity, producing an N mineralization bloom. Using field-moist samples might cause less physical damage during preincubation protocols and cause a better transition from field to lab conditions than dried and/or ground soil samples. However, field-moist soil samples intended for incubation need to be gently crushed through the sieve (usually 2–4 mm) immediately after sample collection.

Homogenized Soil versus Undisturbed Soil Cores

Most laboratory incubation methods utilize a homogenized sample created by sieving. However, there are reports that homogenized samples do not well represent the effects of field soil tillage. Laboratory soil should have a physical structure similar to that of the field environment the sample represents, but sieving artificially "tills" soil from undisturbed/no-tillage environments. This can expose aggregate-protected SOM and enhance the microbial activity, over-estimating the N mineralization. Undisturbed cores may be a better option for laboratory incubations intended to differentiate the impact of tillage on N mineralization.

Constant Temperature versus Variable Temperature

Most laboratory incubation methods use a constant temperature, which does not reflect temperature fluctuation in field conditions. Carpenter-Boggs et al. proposed a variable temperature method for laboratory incubation in which soil samples are incubated in a variable temperature incubator (VTI) that mimicked field soil temperatures under a growing corn canopy. They reported that the VTI technique provided a lower sample variance and a smaller initial flush of N mineralization than constant incubation temperature (35 °C).

Field (in-situ) Incubation Methods

Due to the uncertainty about the extrapolation of laboratory N mineralization values in the field, estimating N mineralization from SOM and crop residues in field conditions would be a compelling research topic for investigators because more efficient N fertilization practices could be hastened if a reliable in-situ N mineralization method was developed. So far, there have been three dominant in-situ research techniques: buried polyethylene bags, covered cylinders, or resin-trap core methods.

Buried Polyethylene Bag Method

The buried polyethylene bag method for in-situ N mineralization was proposed by Eno. The main driving force behind this technical development was the realization that soil temperature variance results in considerable changes in the soil NO_3- production rate. In that preliminary laboratory study, soil in sealed polyethylene bags had an equal nitrification rate compared to soil contained in ventilated bottles. Polyethylene is permeable to oxygen and carbon dioxide, but no NO_3- diffused through the polyethylene bag during the 24-week incubation period. The preliminary results and polyethylene characteristics mean this technique has the potential to estimate aerobic in-situ soil N mineralization.

Although this technique mimics field temperature conditions at a low cost, the technique does not reflect transient field moisture conditions. Elevated NO_3- and carbon dioxide concentrations inside the bags may promote denitrification. Physical damage to the bags by insects or plant roots may result in loss of mineralized N into the field soil via diffusion and mass flow. Another major limitation of this technique is the inevitable disturbance of soil, which does not allow a valid comparison of tillage effects on N mineralization in field conditions.

Covered Cylinder Method

The covered cylinder method was developed as a more durable alternative to the buried bag. This technique allows incubation of intact soil cores. Covered cylinders are usually constructed from PVC or metal pipes that are capped to exclude rainfall, which is also assumed to stop inorganic N leaching. Although the tubes are open at the bottom, aeration is less than that in field soil, which might result in higher denitrification potential. Modifications such as using gas permeable caps or perforations in the tube sidewall were often added to promote air exchange and reduce denitrification potential. However, sidewall aeration holes could potentially allow mineralized N loss. Water may enter the soil tubes through aeration holes, causing N leaching at the bottom of the soil column. Furthermore, plant roots may potentially grow into the soil column via aeration holes or the open bottom, absorbing mineralized soil from the tubes. Another major limitation of this technique is that the soil in the tube usually has a lower soil moisture content than the surrounding field.

The basic principle of the covered cylinder method was to limit N leaching by sheltering incubating soil from precipitation. Based on the same principle, there was another in-situ method called the "rain shelter" which simply used a shelter over the sampled area to prevent leaching. Except for considerations regarding the quality and durability of the rain shelter and surface water run-on during intense rainfall, the major drawback of this technique is a lack of ability to reflect field soil moisture fluctuations.

Resin-trap Soil Core Method

Buried polyethylene bags and covered cylinder methods can capture variation in field temperature, while failing to reflect moisture and aeration conditions, which are reported to play a large role in soil N mineralization. An alternative in-situ method was proposed that employs ion exchange resins to capture mineralized N leaching from undisturbed soil cylinders. The major modification of this technique is an open cylinder top, which allows precipitation and air to freely enter the intact soil column, and a resin trap at the bottom to capture inorganic N that might otherwise leach from the tube. There are some concerns about whether the soil tube causes abiotic differences between soil in the tube and the surrounding field soil. Wienhold et al. reported that soil inside the cylinders was slightly wetter and warmer than adjacent soil, which likely increases soil N mineralization. They pointed out that the magnitude of change in soil N mineralization was likely much less than the normally observed field core-to-core variation. This method was found to better track true field conditions and has the potential to become a standard procedure.

The drawback with intact cores and resin bags is a large resource demand. This technique requires preliminary studies to ensure leached ions are efficiently trapped under field conditions. Resin duality, adsorption capacity, and bypass flow are all factors that can potentially influence resin effectiveness in capturing leached N. The extraction of adsorbed N from the resin is also time-consuming. Kolberg et al. reported that five extractions with KCl were required to completely release adsorbed N.

Other Modifications to In-situ Incubation Methods

Except for the major design developments mentioned above, some minor modifications to in-situ incubation methods have been suggested. Hatch et al. proposed a method to combine the soil core with acetylene inhibition, which would limit N loss by nitrification due to uncontrolled soil in-situ incubation conditions. The big concern with this modification is that the tube must be sealed at the top, causing a loss in practical application to the field environment if rainfall is a concern. Given consideration on different drainage characteristics in resin-trap soil cores relative to the surrounding soil, Hanselman et al. developed a "new" type of resin-trap soil core method in which resin is mixed with soil to create an artificial uniform soil column. This method is impractical when undisturbed soil structure is a research concern, as in a comparison of conventional and conservation tillage.

Method Selection

As discussed above, each method, including laboratory and in-situ methods, has unique assumptions, advantages, and disadvantages. There is no standard method that will work for every situation. The selection of method depends on the nature of the study, available resources, and site-specific factors. Although laboratory methods might not reflect natural field conditions, they can provide reasonable relative values to estimate differences due to soil type and certain management practices. Zou et al. reported that discarding plant residue in the laboratory incubation method neglects the potential effect of plant residue on soil N mineralization. The primary merit to field incubation is a more practical estimation of N mineralization, which might be more useful in management decision making. However, the substantial time and apparatus requirement for the in-situ incubation methods must be considered. Zou et al. reported that soil C and N fractions contribute variably to predict soil N mineralization in different rotation systems, but SOC (which can be calculated from soil organic matter, a common index in the routine test package of many soil testing laboratories) was the best overall NSNM predictor in their study. The principle is that both biotic and abiotic factors control the soil N mineralization process. Knowing the advantage and disadvantage of each method can help the investigator choose the best method while reducing misinterpretation.

Influence of Agronomic Practices on Root Architecture

Plant roots are a fundamental component of terrestrial ecosystems and function to maintain nutrient and water supply to the plant. Although root system architecture is controlled mainly by genetic factors, plant root systems exhibit high developmental plasticity. This plasticity is possible because root development results from continuous propagation of new meristems. In a heterogeneous soil matrix, a wide array of physical, chemical, and biological factors can affect the initiation and activity of root meristems. Previous studies have reported that certain crop root traits enhance productivity in resource-limited environments due to improved nutrient and water scavenging abilities. Agronomic practices can influence crop nutrient uptake capacity by affecting the root growth environment.

Tillage affects root growth mainly by changing soil structure, strength, and penetration resistance. Any particular root increases its length through primary growth when cells of the meristem divide, elongate, and push the root tip forward through surrounding materials. Turgor pressure in the elongating cells is the driving force and must be sufficient to overcome cell wall constraints and other additional constraints imposed by the surrounding environment. Compared to conventional plow tillage, numerous studies on grain crops report that no-tillage increases mechanical impedance, which can result in reduced root length density, root surface density, and lower biomass production. Similar results were found in a no-tillage burley tobacco study. Furthermore, greater mechanical impedance with no-tillage not only restricts root growth but also changes root morphology, restricting main root axis elongation, stimulating lateral root branching and root thickening.

Nutrient supply and distribution (or fertilizer application) can affect root system architecture mainly by signaling. Typically, roots proliferate in volumes where nutrients are most concentrated. However, the mechanisms of plant root response to the different nutrient elements might be controlled by different pathways and signals.

There have been few studies on the effect of crop rotation on plant root architecture. Given the basic factors controlling root development, the hypothesis is crop rotation may differentially influence root architecture compared to monocropping systems, if rotated with residue-rich or deep-rooted crops that can increase SOM levels and soil structure. In this case, rotation affects root proliferation by changing soil structure in a manner similar to that observed with no-tillage. If rotation involves legumes, more N nutrition is provided than found with monocropping. In this case, rotation could affect root architecture by changing soil nutrient supply in a manner similar to that found with fertilizer application.

The effects of agronomic practices on crop N uptake not only affect SOM sequestration and soil N mineralization but also alter the soil environment for plant root proliferation. soil productivity is broadly defined as the soil's unique ability to supply water, nutrients, air, and heat, among other life-sustaining resources, adjusting that supply to the demands of plants and microbes. Soil resources fall into two main components: (a) nutrients and moisture and (b) an environment suited for root growth and microbial activity.

References

- Subsistence_farming, entry: newworldencyclopedia.org, Retrieved 23 June, 2019

- "Companion Planting Guide". Mel's Garden. 2018-07-11. Retrieved 12 July 2018

- Bushnell, G. H. S. (1976). "The Beginning and Growth of Agriculture in Mexico". Philosophical Transactions of the Royal Society of London. 275(936): 117–120. Doi:10.1098/rstb.1976.0074

- Crop-rotation, topic: britannica.com, Retrieved 25 August, 2019

- "5 Secrets to Vegetable Garden, Companion Planting Revealed". Organic Authority. 22 October 2018. Retrieved 1 May 2019

- Bartholomew, Mel (2013). All New Square Foot Gardening (2nd ed.). Cool Springs Press. ISBN 978-1591865483

- Paull, John & Hennig, Benjamin (2016) Atlas of Organics: Four Maps of the World of Organic Agriculture Journal of Organics. 3(1): 25-32

- Agronomic-practices-for-the-management-of-seed-agriculture, agriculture, management: businessmanagementideas.com, Retrieved 4 January, 2019

3

Plant Breeding

The science of changing the characteristics of plants to increase their utility and value for humans is referred to as plant breeding. It is used to increase tolerance towards heat stress and drought stress in plants. This chapter discusses in detail the techniques related to breeding of plants to overcome these stresses as well as the selection methods which are used in this field.

Plant breeding is a method of altering the genetic pattern of plants to increase their value and utility for human welfare. It is a purposeful manipulation of plants to create desired plant types that are better suited for cultivation, give better yield and are disease resistant. Plant breeding is done for the following objectives:

- Increase the crop yield.

- Improve the quality of the crop.

- Increase tolerance to environmental conditions like salinity. extreme temperatures and drought.

- Develop a resistance to pathogens.

- Increase tolerance to the insect pest.

Steps for Different Plant Breeding Methods

The main steps of the plant breeding program are as follows:

Collection of Variability: Wild varieties species and relatives of the cultivated species having desired traits should be collected and preserved. The entire collection having all the diverse alleles for all genes in a given crop is called germplasm collection. Germplasm conservation can be done following ways:

- In situ conservation - It can be done with the help of forests and Natural Reserves.

- Ex situ conservation - it is done through botanical gardens, seed banks.

Evaluation and Selection of Parents: The germplasm collected is evaluated to identify the plants with desirable characters. It is made sure that only the pure lines are selected. The selected plants are multiplied and used in the process of hybridization.

Hybridization: The Pollen Grain from one desired parent plant selected as a male parent is collected and dusted over another plant which is considered as the female parent.

Selection and Testing of Superior Recombinants: Progeny obtained after crossing are evaluated for the desired combination of characters. These are self-pollinated for several generations till there is a state of uniformity so that the characters will not segregate further.

Testing Release and Commercialization of New Cultivars: The selected plants are evaluated by growing the plants in an experimental field and the performance is recorded. This is done for at least 3 growing Seasons at different locations in the country.

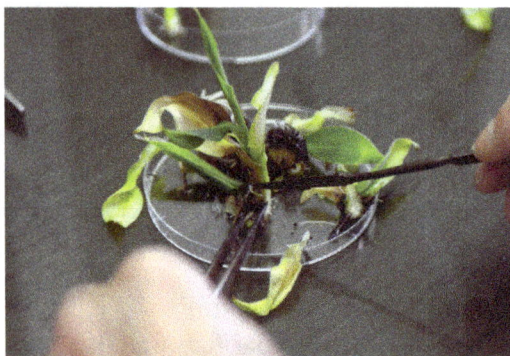

Plant Breeding for Disease Resistance

Some of the diseases caused by plants are:

Fungi	Brown rust of wheat
	Red rot of sugarcane
	Late blight of potato
Bacteria	Black rot of crucifers
Viruses	Tobacco mosaic turnip music

The basic objective of breeding for disease resistance is to develop inherent quality in the plant to prevent the pathogen from causing the disease. Such varieties of plants are called resistant plants. The basic method is the same as normal hybridization. For hybridization resistant plant should be available for breeding.

Some diseases resistant plants developed are:

Crop	Variety	Resistant to Disease
wheat	Himgiri	Leaf & stipe rust
Brassica	Pusa Swarnim	White rust
Cauliflower	Pusa Shubra	Black rot and curl blight
Chilli	Pusa Sadabahar	Chilli mosaic virus

Mutation Breeding

If resistant variety is not available, the resistance can be developed by inducing mutations in the plant through diverse means and then by screening the plant material for resistance.

Mutation is changed in the base sequence of the genes, induced by certain Chemicals or radiations. The resistant plants developed can be multiplied directly or can be used in other breeding experiments.

Plant Breeding for Developing a Resistance to Insect Pest

Resistance can be developed by following ways:

- Development of morphological characters like hairy leaves in cotton and wheat develop vector resistance from jassids beetle.

- Solid stem in wheat lead to resistance from stem borers.

- Biochemical characters provide resistance to insects and pests. For example, the high aspartic acid and low nitrogen and sugar content in maize leads to resistance to maize stem borers.

- Smooth leaves and nectarless cotton develop resistance from bollworms.

Some pest-resistant varieties are:

Crop	Variety	Insect pest
Brassica	Pusa Gaurav	Aphids
Flat bean	Pusa sem	Jassids, aphids
Okra	Pusa Sawni	Shoot and fruit bores

Plant Breeding for Improved Food Quality

Biofortification is a method in which crops are bred for higher levels of vitamins, minerals, and fats. Due to this problem of malnutrition can be overcome. Following objectives were considered for the breeding program:

- Protein content and quality.

- Oil content and quality.

- Vitamin content.

- Micronutrient content and quality.

Some examples of biofortification:

- Fortified Maize having twice the amount of amino acids lysine and tryptophan.

- Atlas 66 wheat has a high protein content.

- Iron-fortified rice having 5 times more iron.

- Vegetable crops like carrot and spinach have more vitamins and minerals.

Issues and Concerns

Modern plant breeding, whether classical or through genetic engineering, comes with issues of concern, particularly with regard to food crops. The question of whether breeding can have a negative effect on nutritional value is central in this respect. Although relatively little direct research in this area has been done, there are scientific indications that, by favoring certain aspects of a plant's development, other aspects may be retarded. A study published in the Journal compared nutritional analysis of vegetables done in 1950 and in 1999, and found substantial decreases in six of 13 nutrients measured, including 6% of protein and 38% of riboflavin. Reductions in calcium, phosphorus, iron and ascorbic acid were also found. The study, conducted at the Biochemical Institute, University of Texas at Austin, concluded in summary: "We suggest that any real declines are generally most easily explained by changes in cultivated varieties between 1950 and 1999, in which there may be trade-offs between yield and nutrient content."

The debate surrounding genetically modified food during the 1990s peaked in 1999 in terms of media coverage and risk perception, and continues today – for example, "Germany has thrown its weight behind a growing European mutiny over genetically modified crops by banning the planting of a widely grown pest-resistant corn variety." The debate encompasses the ecological impact of genetically modified plants, the safety of genetically modified food and concepts used for safety evaluation like substantial equivalence. Such concerns are not new to plant breeding. Most countries have regulatory processes in place to help ensure that new crop varieties entering the marketplace are both safe and meet farmers' needs. Examples include variety registration, seed schemes, regulatory authorizations for GM plants, etc.

Plant breeders' rights is also a major and controversial issue. Today, production of new varieties is dominated by commercial plant breeders, who seek to protect their work and collect royalties through national and international agreements based in intellectual property rights. The range of related issues is complex. In the simplest terms, critics of the increasingly restrictive regulations argue that, through a combination of technical and economic pressures, commercial breeders are reducing biodiversity and significantly constraining individuals (such as farmers) from developing and trading seed on a regional level. Efforts to strengthen breeders' rights, for example, by lengthening periods of variety protection, are ongoing.

When new plant breeds or cultivars are bred, they must be maintained and propagated. Some plants are propagated by asexual means while others are propagated by seeds. Seed propagated cultivars require specific control over seed source and production procedures to maintain the integrity of the plant breeds results. Isolation is necessary to prevent cross contamination with related plants or the mixing of seeds after harvesting. Isolation is normally accomplished by planting distance but in certain crops, plants are enclosed in greenhouses or cages (most commonly used when producing F1 hybrids).

Role of Plant Breeding in Organic Agriculture

Critics of organic agriculture claim it is too low-yielding to be a viable alternative to conventional agriculture. However, part of that poor performance may be the result of growing poorly adapted varieties. It is estimated that over 95% of organic agriculture is based on conventionally adapted varieties, even though the production environments found in organic vs. conventional farming systems are vastly different due to their distinctive management practices. Most notably, organic farmers have fewer inputs available than conventional growers to control their production environments. Breeding varieties specifically adapted to the unique conditions of organic agriculture is critical for this sector to realize its full potential. This requires selection for traits such as:

- Water use efficiency.

- Nutrient use efficiency (particularly nitrogen and phosphorus).

- Weed competitiveness.

- Tolerance of mechanical weed control.

- Pest/disease resistance.

- Early maturity (as a mechanism for avoidance of particular stresses).

- Abiotic stress tolerance (i.e. drought, salinity, etc).

Currently, few breeding programs are directed at organic agriculture and until recently those that did address this sector have generally relied on indirect selection (i.e. selection in conventional environments for traits considered important for organic agriculture). However, because the difference between organic and conventional environments is large, a given genotype may perform very differently in each environment due to an interaction between genes and the environment. If this interaction is severe enough, an important trait required for the organic environment may not be revealed in the conventional environment, which can result in the selection of poorly adapted individuals. To ensure the most adapted varieties are identified, advocates of organic breeding now promote the use of direct selection (i.e. selection in the target environment) for many agronomic traits.

There are many classical and modern breeding techniques that can be utilized for crop improvement in organic agriculture despite the ban on genetically modified organisms. For instance, controlled crosses between individuals allow desirable genetic variation to be recombined and transferred to seed progeny via natural processes. Marker assisted selection can also be employed as a diagnostics tool to facilitate selection of progeny who possess the desired trait(s), greatly speeding up the breeding process. This technique has proven particularly useful for the introgression of resistance genes into new backgrounds, as well as the efficient selection of many resistance genes pyramided into a single individual. Unfortunately, molecular markers are not currently available for many important traits, especially complex ones controlled by many genes.

Addressing Global Food Security through Plant Breeding

For future agriculture to thrive there are necessary changes which must be made in accordance to arising global issues. These issues are arable land, harsh cropping conditions and food security which involves, being able to provide the world population with food containing sufficient nutrients. These crops need to be able to mature in several environments allowing for worldwide access, this is involves issues such as drought tolerance. These global issues are achievable through the process of plant breeding, as it offers the ability to select specific genes allowing the crop to perform at a level which yields the desired results.

Increased Yield without Expansion

With an increasing population, the production of food needs to increase with it. It is estimated that a 70% increase in food production is needed by 2050 in order to meet the Declaration of the World Summit on Food Security. But with the degradation of agricultural land, simply planting more crops is no longer a viable option. New varieties of plants can in some cases be developed through plant breeding that generate an increase of yield without relying on an increase in land area. An example of this can be seen in Asia, where food production per capita has increased twofold. This has been achieved through not only the use of fertilisers, but through the use of better crops that have been specifically designed for the area.

Breeding for Increased Nutritional Value

Plant breeding can contribute to global food security as it is a cost-effective tool for increasing nutritional value of forage and crops. Improvements in nutritional value for forage crops from the use of analytical chemistry and rumen fermentation technology have been recorded since 1960; this science and technology gave breeders the ability to screen thousands of samples within a small amount of time, meaning breeders could identify a high performing hybrid quicker. The main area genetic increases were made was in vitro dry matter digestibility (IVDMD) resulting in 0.7-2.5% increase, at just 1% increase in IVDMD a single Bos Taurus also known as beef cattle reported 3.2% increase in daily gains. This improvement indicates plant breeding is an essential tool in gearing future agriculture to perform at a more advanced level.

Breeding for Tolerance

Plant breeding of hybrid crops has become extremely popular worldwide in an effort to combat the harsh environment. With long periods of drought and lack of water or nitrogen stress tolerance has become a significant part of agriculture. Plant breeders have focused on identifying crops which will ensure crops perform under these conditions; a way to achieve this is finding strains of the crop that is resistance to drought conditions with low nitrogen. It is evident from this that plant breeding is vital for future agriculture to survive as it enables farmers to produce stress resistant crops hence improving food security. In countries that experience harsh winters such as Iceland, Germany and further east in Europe, plant breeders are involved in breeding for tolerance to frost, continuous snow-cover, frost-drought (desiccation from wind and solar radiation under frost) and high moisture levels in soil in winter.

Participatory Plant Breeding

Participatory plant breeding (PPB) is when farmers are involved in a crop improvement programme with opportunities to make decisions and contribute to the research process at different stages. Participatory approaches to crop improvement can also be applied when plant biotechnologies are being used for crop improvement. Local agricultural systems and genetic diversity are developed and strengthened by crop improvement, which participatory crop improvement (PCI) plays a large role. PPB is enhanced by farmers knowledge of the quality required and evaluation of target environment which affects the effectiveness of PPB.

CONVENTIONAL PLANT BREEDING

Plant breeding is the process of selecting plants with the most desirable qualities to produce off-spring that inherit these desired traits. It aims to develop improved crop cultivars – crops selected for desirable characteristics that can be reproduced – to satisfy a variety of needs and overcome a multitude of challenges. Different cultivars are needed for different growing environments, such as rain-fed versus irrigated farming or upland versus paddy rice production. Insect and disease resistant cultivars are often highly desirable, if not essential, especially where pesticides are ineffective or restricted. Breeders also seek to produce crop varieties that appeal to consumer taste, satisfy cooking preferences, adhere to food safety regulations, enhance nutritional quality or behave best when used for industrial applications.

A hybrid rice field.

The application of genetics in crop improvement has yielded many successes leading to unprecedented growth in agricultural productivity, spurred by government commitments to agricultural research and development and supporting sectors. The Green Revolution was the spread of short-strawed, fertiliser-responsive varieties of wheat and rice that led to 'quantum leaps' in food supplies in many Asian countries. Rice yields grew by 32% and wheat by 51%. Without these advances, it is largely recognised that there would be enormous food deficits in the world today. Successes in breeding new varieties of staple crops were smaller in magnitude but no less important in addressing food insecurity in the region. In East and Southern Africa, improved varieties led to growth in both maize production and yields primarily for small-scale, resource-poor farmers.

Improved wheat variety being grown for seed in farmer's field.

Despite these successes, the performance of local varieties of maize, cassava, millet, and other African food staples lags far behind the rest of the world. Harvests per hectare for major crops like maize can be as much as 80% below their potential. Poorly functioning and developed seed systems across Africa are a major reason for this yield gap. Improving food security will therefore require responsive plant breeding programmes and effective seed delivery systems that develop high quality and well-adapted improved seed varieties that meet the needs and preferences of smallholder farmers.

A rapidly growing population is increasing food demand whilst resources such as good quality land, water and soil are becoming scarcer. As fertilisers such as phosphorous and nitrogen become increasingly scarce and expensive, developing plant varieties that can withstand climate variability, lower rainfalls, warmer temperatures, fewer inputs and pest and disease outbreaks is increasingly vital. Plants that use resources more efficiently or require fewer resources altogether improve the sustainability of agricultural as well as urban and forest ecosystems. New crop cultivars must also be bred that improve both soil and human health.

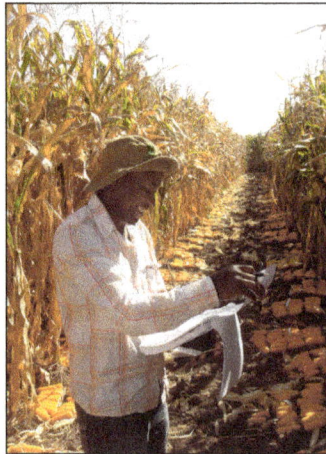

Breeding provitamin A-enriched orange maize.

Conventional plant breeding can occur through a variety of approaches and for a number of objectives, including participatory plant breeding, improving seeds through hybridisation or enhancing their nutritional properties with biofortification. Rather than employing a 'top-down' model whereby seed varieties are improved by professional breeders and then made available to farmers, participatory plant breeding involves farmers in the various stages and decision-making milestones during the breeding process. Hybridisation, the deliberate cross of two genetically different plants from two separate plant populations, tends to create varieties that mature earlier, produce higher yields and greater uniformity. Micronutrient deficiencies, a health problem affecting more than 2 billion people worldwide, can be addressed in part by breeding more nutritive varieties of crops through the process of biofortification.

Hybridisation

Some of the best advances in conventional plant breeding have been in the production of hybrid maize and rice. A cross-pollinating crop, maize seed is naturally a hybrid. Saved hybrid seed is usually highly variable and produces uncertain and often less favourable outcomes. Significant improvements in maize seed breeding were not achieved until the discovery of 'detasseling,' by which the male flowers and their pollen are removed, preventing the plants from fertilising themselves. This paved the way for producing controlled crosses or hybrids that combined the best features of the breeding lines. The process of hybridisation is now more sophisticated and costly to produce hybrid varieties for specific environments. Today, farmers have a choice between open-pollinated varieties (OPVs) (pollination that occurs without deliberate cross-breeding of two separate parent lines) and hybrid varieties, each with advantages and disadvantages. OPVs produced by farmers are usually genetically diverse and not very uniform, but over time the plants become well adapted to their environments. Their costs may be low, but they tend to yield 10% to 25% less than hybrids. In comparison, hybrids are very uniform, but their seeds need to be repurchased each year because their vigour (desired characteristics such as yield, rate of growth and maturity) diminishes with each generation.

Highland yellow single cross hybrid maize ears.   Hybrid rice.

Hybridisation of rice is more complicated because it is a self-pollinated crop and the male and female organs grow on the same floret and only flower for a short time. When the male and female organs do not occur in separate flowers, producing hybrids can be tedious and expensive as a series of delicate, exacting and properly timed operations that must be done by hand are often required. Recently, a built-in cellular system of pollination control known as cytoplasmic male sterility (CMS) has made hybrid varieties of rice and a wide range of other self-pollinating plants, fruits and vegetables possible. Despite advances in hybrid seed technologies, seed systems and breeding programs in Africa remain underdeveloped, underfunded and fragmented, especially when it comes to reaching and meeting the needs of smallholder farmers. From 1966 to 1990, African public maize breeding programs released close to 300 varieties, of which 100 were hybrids. The yield gains were estimated to be 30% in dry areas and 40% in favourable environments, but wide-scale adoption remains low in most countries.

Contribution to Sustainable Intensification

The development of hybrid maize seed has had one of the greatest impacts on increasing the quantity of available food supply. Hybrid rice – grown on more than 21 million hectares, or 13% of land devoted to rice globally – too has contributed substantially to global food production. Hybrid rice increases yields between 5% and 15% over inbred varieties and China's "super" rice achieves yields of 12 tonnes per hectare or 8% to 15% more than other hybrid varieties. Hybrid seeds are usually an improvement over non-hybrids in terms of qualities such as yield, resistance to pests and diseases, and time to maturity. In addition to qualities like good vigour, trueness to type, heavy yields and high uniformity, other characteristics such as earliness, disease and insect resistance and good water holding ability have been incorporated into most hybrids varieties. Hybridisation can also be very useful in developing seed varieties that are drought and pest tolerant, enabling adaptation to climate change or mitigating other risks. As these cultivars have higher yield potential, under the right inclusive markets and climatic conditions, they can contribute to the improvement of farmers' income and asset accumulation as well as the improvement of their standard of living.

Benefits and Limitations

Yield Benefits

The advantage of growing hybrid seed compared to inbred, open-pollinated lines comes from the ability to cross the genetic materials of two different, but related plants to produce new, desirable traits that cannot be produced through inbreeding or selection. Hybridisation can also be very useful in developing seed varieties that are drought and pest tolerant, enabling adaptation to climate change or mitigating other yield penalising risks. Hybrid seed also produces plants that are uniform, and usually higher in yield. Yield benefits vary by crop and environment, but some notable examples include maize hybrids developed by the International Institute for Tropical Agriculture (IITA) in West and Central Africa in the 1980's that by the early 1990's were available with resistance to maize streak virus and out-performed open-pollinated varieties (OPVs) by 33% to 45%.

Sustainable Seed Systems

A sustainable seed system will ensure that high quality seeds of a wide range of varieties and crops are produced and fully available in time, acceptable and affordable to farmers. However, farmers are not always able to fully benefit from the advantages of using improved seed. Part of these challenges will not be addressed without the development and improved access to finance and markets, but reforms and improvements are also needed within research and breeding programmes. Seed systems are usually divided into formal and informal; the former consisting of the development of new varieties by trained plant breeders and multiplied according to set standards and the latter informal seed selection, multiplication and exchange by farmers. Whilst the formal sector offers opportunities to create uniform, quality and reliable seed that offer greater yield or other desired benefits, these seeds can sometimes be cost-prohibitive. Informal systems are low cost and adaptable to farmer preferences and needs, and the use of multiple seed varieties on a farm can improve resilience to certain to shocks and stresses, but is limited in its ability to develop or widely disseminate new varieties. As both approaches offer options for strengthening resilience against a variety of risks and uncertainties, and private seed companies have not demonstrated strong interest in producing seeds that are grown in small volumes, in remote areas or marginal areas, both systems should be strengthened.

Maize as an Example

Maize is the most important cereal crop in sub-Saharan Africa. Africa produces 6.5% of maize globally, but accounts for 30% of worldwide consumption. Eastern and Southern Africa consumes 85% of its production for food, whilst Africa as a whole consumes 95%, compared to other regions of the world that use maize primarily as animal feed. Although many other crops continue to be culturally important, offer more nutrients or strategies for diversification and resilience, maize receives the majority of attention from agricultural research and breeding programmes. Recently, however, there has been growing attention to breeding other cereals such as millet and sorghum.

Plant breeding has large fixed costs and long (5-20 years) and uncertain payoffs. Typically, only a large company or a subsidised public entity can afford to assume these risks. In addition, many types of seed are easy to reproduce. Though, some seeds are difficult to harvest or store such as many vegetable seeds, others do not maintain their yield advantage if saved, as is the case with hybrid maize. Many self-pollinating crops including wheat, rice, groundnuts, and potatoes maintain their genetic characteristics over many generations. This explains why farmers generally will purchase vegetable and hybrid maize seed, but often use saved seed for other crops. In addition to the difficulty for a company to recover its breeding costs when farmers save and re-use seed, limited consumer demand for other types of seed varieties from the formal sector also constrains investments in neglected crops and environments.

Adoption

In a number of countries, the use of hybrid cultivars in sorghum and maize production has been increasing, for example in Nigeria, Kenya and Zambia. Rates of adoption, however, vary by country, region and farm typology and size. In Kenya, hybrid seed is more widely adopted by commercial rather than small-scale farmers. The use of hybrid technology is also considerably higher in the South African commercial farming sector than in the smallholder sector. In contrast, hybrid cultivars dominate agricultural food production systems of small-scale farming sectors in Zambia and Zimbabwe. One study noted that hybrid technology is likely to be adopted more extensively by farmers that have large farms. Farmers who have access to farming support services also tend to swiftly adopt new technology. In Kenya, for example, farmers' access to credit, input supply and extension services enhanced the adoption of their new hybrid maize. Another study found that gender and literacy levels affect adoption of new seed technologies. Although women constitute the majority of the subsistence farming community, men are more likely to adopt hybrid sorghum. Farmers with some level of education adopt the hybrid cultivars more than those without years of schooling. However, in other studies this has not been found to be a strong predictor of use. Distance to roads, access to transport infrastructure and farm size are found to influence the scale of hybrid seed planted by smallholder maize growers most significantly. Further, the majority of farmers with membership in cooperatives use hybrid sorghum. Therefore, access to agricultural organisations may also have an impact on the farmer adoption of technology. In Kenya, cost was cited as the strongest barrier to hybrid seed adoption.

Economic Impacts

The advantages of hybrids come at a price, because the performance of seeds diminishes in subsequent generations, requiring annual purchasing to maintain performance. Despite this reoccurring cost, the influence of hybrid seed on income and assets is favourable for smallholder maize growers in Kenya and Zambia. Although smallholders in Kenya, particularly in less favourable areas, have been reluctant to adopt hybrid maize varieties, well adapted varieties can be profitable in terms of yield and yield stability even in marginal areas under low input conditions. Further, those that do not adopt the technology, can be considered disadvantaged. In Zambia, the use of maize hybrids is associated with higher values of household income, assets, farm and processing equipment, livestock, and lower levels of deprivation compared to non-adopting farmers. For every kilogram of hybrid maize seed planted, total household income increased by ZMK 32,230 (US$6,000) on average. Since this is considerably more than the price of hybrid seed in Lusaka

ZMK 4,300 – 12,000 (US$0.59 – 1.65), hybrid maize proved profitable for Zambian smallholder farmers. If this profitability continues, returns will build productive assets and lessen the severity of poverty.

PARTICIPATORY PLANT BREEDING

Participatory plant breeding (PPB) is when farmers are involved in a plant breeding programme with opportunities to make decisions at different stages during the process. Farmer's involvement in PPB can include defining breeding goals and priorities, selecting or providing germplasm, hosting trials in their own fields, selecting superior plants for further breeding, engagement in the research design and administration processes as well as the commercialisation of selected lines.

Researchers and farmers select improved finger millet varieties together.

PPB was developed as an alternative and complementary breeding approach to conventional plant breeding. Conventional plant breeding is generally carried out by trained breeders in laboratory or controlled environments, often under favourable farming conditions. The main objectives of conventional breeding programmes tend to focus on 'broad adaptability' or the capacity of a variety to produce high yields over a range of environments and years. Conversely, PPB involves breeders, farmers and other 'consumers' or end users such as rural farm associations or cooperatives in plant breeding research. This enables breeders to better understand the local farming conditions, the farmers' traditional ways for managing plant diversity as well as their specific needs and preferences. The aims of PPB are also more targeted, focusing on breeding for individual environments and needs. In fact, most progress with PPB has been in marginal or neglected environments (those that are naturally harsh climates or are excluded from connective infrastructure such as roads and markets).

Improving seed through on-farm mass selection by farmers was the 'conventional' method for varietal improvement until the process was 'sped-up' under the direction of large-scale breeding programmes primarily in developed countries during the 20th century. Crop improvement during the Green Revolution invested in varietal development and packages of inputs that were considered economically optimal. Dissatisfaction with the ability of these 'optimal production packages' to meet the needs of smallholders in difficult growing environments led to the establishment of local-breeding programmes. Since advocated by NGO's and also undertaken by some public research institutions, there is now general consensus that farmers need to be more involved in the breeding process.

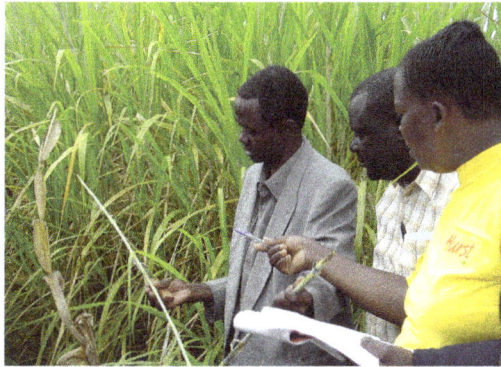
Farmer and researchers assess crops.

Within marginal areas, PPB has demonstrated a number of advantages and positive impacts, but the evidence base for drawing strong conclusions for its success is thin. Where PPB has resulted in higher adoption rates, this is attributed to greater farmer ownership over the breeding process and pre-assurance that the varieties are suitable to the needs of the farmers and their communities. PPB can also reduce the lag-time between variety testing and release helping to get farmers access to improved varieties more quickly and encourage the development of diverse, locally adapted plant populations or in situ (on farm) conservation that contributes to improved resilience. Further, PPB can empower groups, especially women or less well-off farmers that are traditionally left out of the development process. Compared to participatory varietal selection (PVS) where farmers are only involved in selecting varieties from amongst a pre-determined group of lines that are being field tested, PPB is found to have a higher empowerment effect.

Contribution to Sustainable Intensification

Many smallholder farmers in developing countries work lands in marginal areas that are either neglected by markets and infrastructure or face harsh growing conditions, or both. Smallholder farmers in developing countries are also one of the groups most vulnerable to the adverse impacts of climate change. Improved crop varieties that are well suited to diverse and individual growing environments, meet both the physical and social preferences of smallholders, and by preserving crop diversity can help to improve food and nutrition security. Greater genetic diversity and the enhancement of crops suited to dry environments can also enable smallholder farmers to increase their resilience and adapt to climate change. Inclusion of farmers in the decision-making and breeding process also strengthens the empowerment of both farmers and often excluded members of communities such as women and less well-off farming households. Participatory plant breeding helps to build stronger and more reliable seed systems for smallholder farmers working in diverse and harsh environments, thus contributing to sustainable food security and strengthened resilience to a variety of risks and challenges.

Improving Adoption Rates

Participatory plant breeding (PPB) makes use of the traditional knowledge of the farmers involved. Valuing traditional knowledge creates incentives for using and developing its by-products. As test fields are typically located on a portion of the farmers' fields, the new varieties can be adapted to real production conditions. By allowing farmers to select varieties suitable for their local soil qualities, rainfall conditions and resource requirements, PPB can lead to crops that not only perform well in their intended environments, but are available more quickly and are more readily adopted. According to the World Bank, PPB halved the time for varietal development and dissemination from 10-15 years to 5–7 years compared to a conventional plant breeding.

Slow Rates of Adoption

Despite the advantages of participatory plant breeding (PPB), regular or long-term adoption in national and international breeding programmes has been slow. The reasons for slow adoption are not well examined, but some explanations include: PPB is best suited to cereals and less-so to roots, tubers and pulses; the general attitude amongst scientists, professionals and decision-makers is that PPB is opposed to conventional plant breeding programmes rather than complementary; and strict criteria for seed registration in many countries can be a disincentive to both public and private breeding programmes that ultimately seek to recover research costs through release and commercialisation. In order to be a sustainable approach to plant breeding, PPB needs to become a permanent feature of plant breeding programmes, addressing crops grown in agriculturally difficult and climatically challenging environments.

Improving Seed Systems

More effective systems for producing and disseminating quality seeds of improved varieties in Africa are necessary in order to reach more farmers with varieties suitable to their growing environments, preferences and resource constraints. Participatory plant breeding (PPB) can strengthen farmer seed systems, defined as the ways in which farmers produce, select, save and acquire seeds. Encouraging and supporting on-farm seed production by farmers is seen as one approach to sustainable seed delivery in Africa because it gives farmers betters access to quality seed of their choice.

Increased Genetic Diversity

Participatory plant breeding (PPB) supports the development and maintenance of a more genetically diverse portfolio of varieties. Unlike the current global breeding model, which for the most part has concentrated on developing a limited number of varieties that are stable over time and adapted to a wide range of environments, the breeding model based on PPB methods encourages the maintenance of more diverse, locally adapted crops. Precisely for this reason, PPB does not attract any significant levels of investment as varieties developed for niche environments are unlikely to spread or spill-over into other areas. Scaling up PPB methods for work at the regional, national, or international level could require large investments in resources.

Costs and Benefits

As farmers' participation increases, they must invest increasing amounts of time, energy, and resources; they must also provide increasing amounts of intellectual input and draw on increasingly sophisticated analytical skills. Whilst this can be taxing or unattractive to farmers, for research institutions, participatory plant breeding (PPB) can be less costly to conduct than traditional breeding. This is due to potential savings on field testing sites, lower overhead costs and the shortening of the research period, by half in some cases, required for producing useful materials. One study reports that "on-farm crop variety evaluations revealed a cost of $0.50 per recorded data point for participatory trials, compared with $0.80 for conventional trials." Anther cost-benefit analysis of participatory and conventional plant breeding conducted in Syria showed that the benefit/cost ratio of PPB is 2.6 times higher than that of conventional plant breeding.

SELECTION METHODS IN PLANT BREEDING BASED ON MODE OF REPRODUCTION

Plant breeders use different methods depending on the mode of reproduction of crops, which include:

- Self-fertilization, where pollen from a plant will fertilise reproductive cells or ovules of the same plant.

- Cross-pollination, where pollen from one plant can only fertilize a different plant.

- Asexual propagation (e.g. runners from strawberry plants) where the new plant is genetically identical to its parent.

- Apomixis (self-cloning), where seeds are produced asexually and the new plant is genetically identical to its parent.

The mode of reproduction of a crop determines its genetic composition, which, in turn, is the deciding factor to develop suitable breeding and selection methods. Knowledge of mode of reproduction is also essential for its artificial manipulation to breed improved types. Only those breeding and selection methods are suitable for a crop which does not interfere with its natural state or ensure the maintenance of such a state. It is due to such reasons that imposition of self-fertilization on cross-pollinating crops leads to drastic reduction in their performance.

For teaching purpose, plant breeding is presented as four categories: Line breeding (autogamous crops), population breeding (allogamous crops), hybrid breeding (mostly allogamous crops, some autogamous crops), clone breeding (vegetatively propagated crops).

Self Fertilizing Crops (Autogamous Crops)

Certain restrictions caused the mechanisms for self-fertilization (partial and full self-fertilization) to develop in a number of plant species. Some of the reasons why a self-fertilizing method of reproduction is so effective are the efficacy of reproduction, as well as decreasing genetic variation and thus the fixation of highly adapted genotypes. Almost no inbreeding depression occurs in self-fertilizing plants because the mode of reproduction allows natural selection to take place in wild populations of such plants.

Critical steps in the improvement of self-fertilizing crops are the choice of parents and the identification of the best plants in segregating generations. The breeder should also have definite goals with the choice of parents. Self-fertilizing are easier to maintain, but this could lead to misuse of seed.

Some of the agronomy important, self-fertilizing crops include wheat, rice, barley, dry beans, soy beans, peanuts, tomatoes, etc.

Mass Selection

This method of selection depends mainly on the selection of plants according to their phenotype and performance. The seed from selected plants is bulked for the next generation. This method is used to improve the overall population by positive or negative mass selection. Mass selection is only applied to a limited degree in self-fertilizing plants and is an effective method for the improvement of land races. This method of selection will only be effective for highly heritable traits. One shortage of mass selection is the large influence that the environment has on the development, phenotype, and performance of single plants. This can also be an advantage in that varieties can be selected for local performance.

Stratified mass selection for ear size over 22 cycles has drastically altered plant phenotype in the maize population Zacatecas 58. Plants in the C22 cycle were 50 cm taller had twice the leaf area index, reached anthesis 7 days later and had a 30% higher harvest index than C0 Differences in growth were detected early in ontogeny. The root growth of C22 exceeded that of C0 and the ratio of shoot dry mass to root dry mass was reduced by nearly 12%, from 8.0 ± 0.2 to 7.1 ± 0.1. Analysis of yield components revealed that C22 was superior to C0 in grain weight, number of rows per ear, number of grains per row, and total yield per unit area. Because the two genotypes were phenologically different, planting density optima are probably different for each population.

Selection of Cross-pollinated Crops

Plant species where normal mode of seed set is through a high degree of cross-pollination have characteristic reproductive features and population structure. Existence of self-sterility, self-incompatibility, imperfect flowers, and mechanical obstructions make the plant dependent upon foreign pollen for normal seed set. Each plant receives a blend of pollen from a large number of individuals each having different genotypes. Such populations are characterized by a high degree of heterozygosity with tremendous free and potential genetic variation, which is maintained in a steady state by free gene flow among individuals within the populations.

In the development of hybrid varieties, the aim is to identify the most productive heterozygote from the population, which then is produced with the exclusion of other members of the population.

Mass Selection

It is the simplest, easiest and oldest method of selection where individual plants are selected based on their phenotypic performance, and bulk seed selection proved to be quite effective in maize improvement at the initial stages but its efficacy especially for improvement of yield, soon came under severe criticism that culminated in the refinement of the method of mass selection. The selection after pollination does not provide any control over the pollen parent as result of which effective selection is limited only to female parents. The heritability estimates are reduced by half, since only parents are used to harvest seed whereas the pollen source is not known after the cross pollination has taken place.

Recurrent Selection

This type of selection is a refined version of the mass selection procedure and differs as follows:

- Visually selected individuals out of the base population undergo progeny testing.

- Individuals selected on basis of the progeny test data are crossed with each other in every possible way to produce seed to form the new base population.

Half-sib Selection with Progeny Testing

Selections are made based on progeny test performance instead of phenotypic appearance of the parental plants. Seed from selected half-sibs, which have been pollinated by random pollen from the population (meaning that only the female parent is known and selected, hence the term "half-sib") is grown in unreplicated progeny rows for the purpose of selection. A part of the seed is planted to determine the yielding ability, or breeding value, for any character of each plant. The seed from the most productive rows or remnant seed from the outstanding half-sibs is bulked to complete one cycle of selection.

Full-sib Selection with Progeny Testing

A number of full-sib families, each produced by making crosses between the two plants from the base population are evaluated in replicated trials. A part of each full-sib family is saved for recombination. Based on evaluation the remnant seed of selected full-sib families is used to recombine the best families.

Breeding of Asexually Propagated Crops

Asexual reproduction covers all those modes of multiplication of plants where normal gamete formation and fertilization does not take place making these distinctly different from normal seed production crops. In the absence of sexual reproduction, the genetic composition of plant material being multiplied remains essentially the same as its source plant.

Clones of mother plants can be made with the exact genetic composition of the mother plant. Superior plants are selected and propagated vegetatively; the vegetative propagated offspring are used to develop stable varieties without any deterioration due to segregation of gene combinations. This unique characteristic of asexual reproduction helped to develop a number of cultivars of fruits and vegetables including grapes, apples, pears and peaches.

Improving Asexual Plant Material through Selection

The selection in these crops is restricted to the material introduced from other sources, such as field plantations. The improvement of asexually propagated plants through induced mutations has distinct advantages and limitations. Any vegetative propagule can be treated with mutagens and even a single desirable mutant or a part of a mutated propagule (chimera) can be multiplied as an improved type of the original variety.

Selection of Asexual Plants

Selection, in the case of asexual plants, can be defined as the selection of the best performing plant and the vegetative propagation thereof. Because plants are not totally genetically stable, it can be expected that deviations would occur through the years. Selection is thus an ongoing process where deviants are selected or removed from the selection program. The main purpose of selection is to better the quality and yield of forthcoming plantations. Different approaches can be followed in the selection process of asexual plants, such as mass selection and clone selection from clone blocks.

In mass selection there are some factors that must be considered when selecting plants in a mother block, e.g. vineyard. Time of selection is a big factor, because you have to select when most of the characteristics of the plant are clearly showing. With asexual perennials the best time is just before harvest. For the best results the selected plant must be evaluated during the next season, when growth-abnormalities, leave disfigurations and virus symptoms are best visualized. Mass selection is done annually on the same plant for a minimum of three years. A plant that does not conform to the requirements in any given year of the selection cycle is discarded from the program.

New Clone Development

The development and registration of new clones take place by means of local clone selection in old plantations, as well as the importation of high quality clones from abroad, for local evaluation.

A clone is the vegetative offspring of one specific mother plant; it does not show any genetic, morphologic or physiologic deviations from the mother plant. Evaluation takes place with the different selected clones after selection.

MOLECULAR PLANT BREEDING

Plant breeding describes methods for the creation, selection, and fixation of superior plant phenotypes in the development of improved cultivars suited to needs of farmers and consumers. Primary goals of plant breeding with agricultural and horticultural crops have typically aimed at improved yields, nutritional qualities, and other traits of commercial value. The plant breeding paradigm has been enormously successful on a global scale, with such examples as the development of hybrid maize, the introduction of wheat and rice varieties that spawned the Green Revolution, and the recent commercialization of transgenic crops. These and many other products of plant breeding have contributed to the numerous benefits global society has received from greater sustainable supplies of carbon that may be harvested as food, feed, forests, fiber, and fuel.

Plant breeding has a long history of integrating the latest innovations in biology and genetics to enhance crop improvement. Prehistoric selection for visible phenotypes that facilitated harvest and increased productivity led to the domestication of the first crop varieties and can be considered the earliest examples of biotechnology. Darwin outlined the scientific principles of hybridization and selection, and Mendel defined the fundamental association between genotype and phenotype, discoveries that enabled a scientific approach to plant breeding at the beginning of the 20th century. Despite the immediate recognition among some plant breeders of the importance of Mendelian genetics, full integration was delayed for nearly 20 years until quantitative genetics reconciled Mendelian principles with the continuous variation observed for most traits considered important by most plant breeders. Subsequent advances in our understanding of plant biology, the analysis and induction of genetic variation, cytogenetics, quantitative genetics, molecular biology, biotechnology, and, most recently, genomics have been successively applied to further increase the scientific base and its application to the plant breeding process.

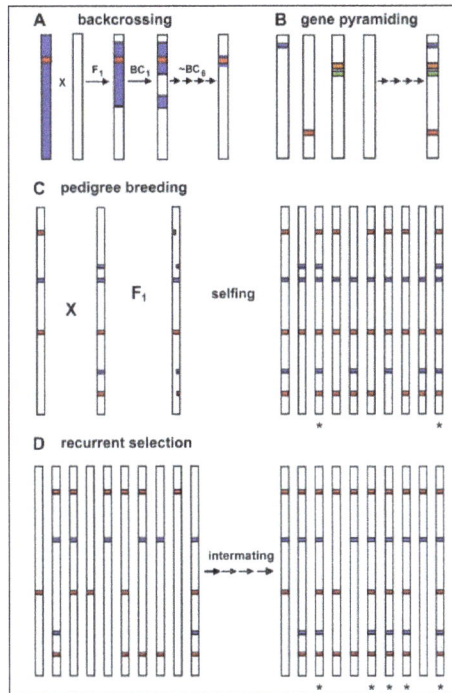

Common breeding and selection schemes. Each vertical bar is a graphical representation of the genome for an individual within a breeding population, with colored segments indicating genes and/or QTLs that influence traits under selection. Genes associated with different traits are shown in different colors (e.g. red, blue). "X" indicates a cross between parents, and arrows depict successive crosses of the same type. Asterisk below an individual signifies a desirable genotype. A, Backcrossing. A donor line (blue bar) featuring a specific gene of interest (red) is crossed to an elite line targeted for improvement (white bar), with progeny repeatedly backcrossed to the elite line. Each backcross cycle involves selection for the gene of interest and recovery of increased proportion of elite line genome. B, Gene pyramiding. Genes/QTLs associated with different beneficial traits (blue, red, orange, green) are combined into the same genotype via crossing and selection. C, Pedigree breeding. Two individuals with desirable and complementary phenotypes are crossed; F1 progeny are self-pollinated to fix new, improved genotype combinations. D, Recurrent selection. A population of individuals (10 in this example) segregate for two traits (red, blue), each of which is influenced by two major favorable QTLs. Intermating among individuals and selection for desirable phenotypes/genotypes increases the frequencies of favorable alleles at each locus. For this example, no individual in the initial population had all of the favorable alleles, but after recurrent selection half of the population possesses the desired genotype. For hybridized crops, recurrent selection can be performed in parallel within two complementary populations to derive lines that are then crossed to form hybrids; this method is called reciprocal recurrent selection.

The plant biotechnology era began in the early 1980s with the landmark reports of producing transgenic plants using Agrobacterium. Molecular marker systems for crop plants were developed soon thereafter to create high-resolution genetic maps and exploit genetic linkage between markers and important crop traits. By 1996, the commercialization of transgenic crops demonstrated the successful integration of biotechnology into plant breeding and crop improvement programs. As depicted in figure, introgression of one or a few genes into a current elite cultivar via backcrossing is a common plant breeding practice. Methods for marker-assisted backcrossing were developed rapidly for the introgression of transgenic traits and reduction of linkage drag, where molecular markers were used in genome scans to select those individuals that contained both the transgene and the greatest proportion of favorable alleles from the recurrent parent genome. During the past 25 years, the continued development and application of plant biotechnology, molecular markers, and genomics has established new tools for the creation, analysis, and manipulation of genetic variation and the development of improved cultivars. Molecular breeding is currently standard practice in many crops, with the following sections briefly reviewing how molecular information and genetic engineering positively impacts the plant breeding paradigm.

Principles and Practices of Molecular Plant Breeding

Breeding Schemes and the Genetic Gain Concept

Conceptually, plant breeding is simple: cross the best parents, and identify and recover progeny that outperform the parents. In practice, plant breeding is a three step process, wherein populations or germplasm collections with useful genetic variation are created or assembled, individuals with superior phenotypes are identified, and improved cultivars are developed from selected individuals. A wide diversity of approaches, tailored to the crop species and breeding objectives, have been developed for improving cultivars. These breeding methods feature different types of populations, selection procedures, and outcomes.

Figure summarizes the three breeding methods that are commonly employed in crop improvement programs. As mentioned previously, when the goal is to upgrade an established elite genotype with trait(s) controlled by one or a few loci, backcrossing is used either to introgress a single gene or to pyramid a few genes. For genetically complex traits, germplasm improvement instead requires reshuffling of the genome to produce new favorable gene combinations in the progeny. The pedigree breeding method produces such novelty via crossing and recombination among superior, yet complementary, parents and selection among segregating progeny for improved performance. Recurrent selection aims to simultaneously increase the frequencies of favorable alleles at multiple loci in breeding populations through intermating of selected individuals. For hybridized crops such as maize, recurrent selection may be extended to improve the performance of distinct complementary populations (e.g. heterotic groups) that are used as parents to form superior hybrid combinations. This practice is referred to as reciprocal recurrent selection.

Quantitative genetic principles have been particularly powerful as the theoretical basis for both population improvement and methods of selecting and stabilizing desirable genotypes (Hallauer, 2007). An important concept in quantitative genetics and plant breeding is genetic gain (ΔG), which is the predicted change in the mean value of a trait within a population that occurs with selection. Regardless of species, the trait of interest, or the breeding methods employed, ΔG serves as a simple universal expression for expected genetic improvement. Figure shows the genetic gain equation and an expansion of its terms to fundamental parameters of quantitative genetics. Though clearly an oversimplification of the advanced quantitative genetic principles employed in plant breeding, the genetic gain equation effectively relates the four core factors that influence breeding progress: the degree of phenotypic variation present in the population (represented by its sd, σP), the probability that a trait phenotype will be transmitted from parent to offspring (heritability, $h2$), the proportion of the population selected as parents for the next generation (selection intensity, i, expressed in units of sd from the mean), and the length of time necessary to complete a cycle of selection (L). L is not only a function of how many generations are required to complete a selection cycle, but also how quickly the generations can be completed and how many generations can be completed per year.

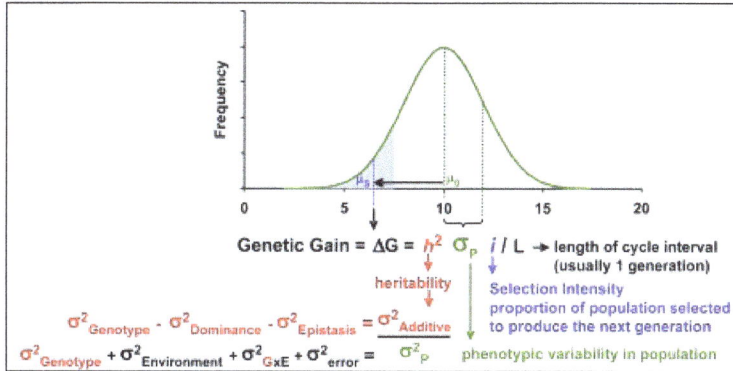

The genetic gain equation and its component variables. The top portion illustrates an idealized distribution showing the frequency of individuals within a breeding population (y axis) that exhibit various classes of phenotypic values (x axis). Mean phenotypic value (μ_0) of the original population (shown as entire area under the normal curve) and mean (μ_S) for the group of selected individuals (shaded in blue) are indicated. In this generalized example, trait improvement is achieved by selecting for a lower phenotypic value, e.g. grain moisture at harvest in maize. Components of variation (σ^2) that contribute to the sd of the phenotypic distribution (σP) are indicated below the histogram.

It is clear that ΔG can be enhanced by increasing σP, h^2, or i, and by decreasing L. Thus, the genetic gain equation provides a framework for comparing the predicted effectiveness of particular breeding strategies and is often used as a guide to the judicious allocation of resources for achieving breeding objectives. When considered in the context of the genetic gain concept, molecular plant breeding offers powerful new approaches to overcome previous limitations in maximizing ΔG. The following sections cite examples where molecular plant breeding positively impacts ΔG and each of its component variables. For brevity, we focus on examples from maize where molecular breeding is most advanced, and has now become the primary means to develop improved commercial hybrids.

Molecular Plant Breeding Expands useful Genetic Diversity

The maximum potential for genetic gain is proportional to the phenotypic variation (σP) present in the original source population and maintained in subsequent cycles of selection. Phenotypic variation is positively associated with genetic diversity, yet also depends on environmental factors and the interactions between genotype and environment. Genetic diversity may be derived from breeding populations (either naturally occurring or synthetic), segregating progeny from a cross of selected parental lines, exotic materials that are not adapted to the target environment, wide interspecific crosses, naturally occurring or induced mutations, the introduction of transgenic events, or combinations of these sources.

However, not all phenotypic variation is equal. For example, the use of exotic germplasm has been extremely successful for improving many crop species, but difficulties may be encountered through the introduction of undesirable alleles associated with lack of adaptation. The need for genetic diversity must be balanced by elite performance, because choosing the best parents is key to maximizing the probability for successful improvement. In contrast, the expected increase in linkage disequilibrium among elite populations derived from intense prior selection may also limit the creation of new genetic combinations for future gain. Intermating source populations for genetic recombination may overcome this problem, but delays cultivar development.

Molecular markers and more recently, high-throughput genome sequencing efforts, have dramatically increased knowledge of and ability to characterize genetic diversity in the germplasm pool for essentially any crop species. Using maize as one example, surveys of molecular marker alleles and nucleotide sequence variation have provided basic information about genetic diversity before and after domestication from its wild ancestor teosinte, among geographically distributed landraces, and within historically elite germplasm. This information enriches investigations of plant evolution and comparative genomics, contributes to our understanding of population structure, provides empirical measures of genetic responses to selection, and also serves to identify and maintain reservoirs of genetic variability for future mining of beneficial alleles. In addition, knowledge of genetic relationships among germplasm sources may guide choice of parents for production of hybrids or improved populations.

While molecular markers and other genomic applications have been highly successful in characterizing existing genetic variation within species, plant biotechnology generates new genetic diversity that often extends beyond species boundaries. Biotechnology enables access to genes heretofore not available through crossing and creates an essentially infinite pool of novel genetic variation. Genes may be acquired from existing genomes spanning all kingdoms of life, or designed and assembled de novo in the laboratory. Both subtle and extreme examples of the power of transgenes to introduce novel phenotypic variation can be found in the three different transgenes developed for resistance to glyphosate herbicides in maize and other crops. The first glyphosate-tolerant maize hybrids used a modified version of the endogenous maize gene encoding 5-enol-pyruvyl-shikimate-3-P synthase, which was followed later by events produced with a 5-enol-pyruvylshikimate-3-P synthase gene isolated from Agrobacterium. More recently, a synthetic gene with enhanced glyphosate acetyltransferase activity was created via gene shuffling and selection in a microbial system. Each of these glyphosate-tolerant maize events also illustrates another benefit of biotechnology, where new combinations of regulatory sequences (e.g. the cauliflower mosaic virus 35S and rice actin1 promoters) may be used to achieve optimal trait expression with respect to overall activity and tissue distribution relative to what might be possible with endogenous genes.

Molecular Plant Breeding Increases Favorable Gene Action

Quantitative genetics uses the theoretical concept of heritability to quantify the proportion of phenotypic variation that is controlled by genotype. In practice, heritability is greatly influenced by the genetic architecture of the trait of interest, which is described by the number of genes, the magnitude of their effects, and the type of gene action associated with phenotypes. Better knowledge of genetic architecture and favorable gene action (that which is more amenable to selection) often has the greatest impact on improving genetic gain. For the genetic gain formula, heritability (h^2) is used in its narrow sense, representing the proportion of phenotypic variation due to additive genetic effects (those that reflect changes in allele dosage or allelic substitutions). Additive genetic effects are also referred to as the breeding value because they are predictably transmitted to progeny. Deviations from additive effects are significant for many traits, and are partitioned into either dominance effects that reflect the interactions between different alleles at the same locus or epistatic effects resulting from interactions among different loci. Gene action and breeding values are characterized by progeny testing, where the phenotypes of individuals in a population are compared to their parents and siblings produced from either self-pollination or outcrossing.

Previous efforts to develop large numbers of molecular markers, high density genetic maps, and appropriately structured mapping populations have now made routine for many crop species the ability to simultaneously define gene action and breeding value at hundreds and often thousands of loci distributed relatively uniformly across entire genomes. The results from such mapping studies provide greatly improved estimates for the number of loci, allelic effects, and gene action controlling traits of interest. More importantly, genomic segments can be readily identified that show statistically significant associations with quantitative traits (quantitative trait loci [QTLs]). In addition to genetic mapping in families derived from biparental crosses, new advances in association genetics with candidate genes and approaches that combine linkage disequilibrium analysis in families and populations further enhance power for QTL discovery.

Information about QTLs can be used in a number of ways to increase heritability and favorable gene action. For traits exhibiting low to moderate heritability, such as grain yield, QTLs, and their associated molecular markers often account for a greater proportion of the additive genetic effects than the phenotype alone. Furthermore, knowledge of genetic architecture can be exploited to add or delete specific alleles that contribute to breeding value. When either genetic linkage or epistasis among loci with antagonistic effects on a trait limits genetic gain, QTL information can be used to break these undesirable allelic relationships.

Success in using information about QTLs to increase genetic gain depends greatly on the magnitude of QTL effects, precise estimation of QTL positions, stability of QTL effects across multiple environments, and whether QTLs are robust across relevant breeding germplasm. Prediction of QTL positions is enhanced by further fine mapping, which facilitates testing QTL effects and breeding values in additional populations. When the density of observed recombinations approaches the resolution of single genes, the causal genetic change for a QTL can be determined. Molecular isolation of QTLs permits the development of perfect or functional molecular markers at the potential resolution of the fundamental unit of inheritance, the nucleotide, and dramatically increases the specificity and precision by which genetic effects are estimated and manipulated in breeding programs.

The use of transgenes can further simplify the genetic architecture for desirable traits, in ways that may be superior to or not possible even when perfect markers are available for robust QTLs of large effect. Transgenes typically condition strong genetic effects at operationally single loci, which also exhibit dominant gene action where only one copy of the event is needed for maximal trait expression in a hybrid cultivar. These features of transgenes can reduce complex quantitative improvement to a straightforward, often dramatic, solution. Excellent examples are provided by the expression in transgenic corn hybrids of insecticidal toxin proteins from Bacillus thuringiensis (Bt) to reduce feeding damage by larvae of the European corn borer (Ostrinia nubilalis) or the corn rootworm beetle (Diabrotica spp.). Partial resistance in maize germplasm to these insect pests had been previously characterized as quantitatively inherited traits with low heritability, but the Bt transgenic events offer a simply inherited alternative that is efficiently manipulated in breeding programs.

By simplifying genetic architecture, transgenes may also permit disruption of allelic interactions between factors controlling the trait of interest and other important performance characteristics. For example, employing a transgenic source of insect resistance (e.g. a single locus Bt transgene) may facilitate selection for favorable alleles for yield improvement that are tightly linked in repulsion with endogenous genes for resistance to the same class of insect pests. In addition, transgenic events may be engineered to uncouple negative pleiotropic effects from beneficial phenotypes conditioned by recessive mutations. This application is illustrated by the use of RNA interference to specifically down-regulate zein seed storage protein gene expression. This strategy mimics the effects of the opaque2 mutation on improving the amino acid profile of maize grain for animal feed, while circumventing the softer endosperm texture and susceptibility to fungal pathogens typically associated with opaque2.

Transgenic events can also be designed to intervene at key regulatory steps for entire metabolic or developmental pathways, such that gene action for the corresponding traits are largely inherited as single dominant factors that are less sensitive to environmental effects. Examples include the expression of a transcription factor that increases drought tolerance, and altering the balance between levels of the GLOSSY15 transcription factor relative to its repressor, microRNA172, to delay flowering time in maize hybrids.

Biotechnology also facilitates the molecular stacking of transgenes that control a trait or suite of traits into a single locus haplotype defined by a transgenic event. Examples of such an approach include the initial Golden Rice, recently released YieldGuard VT Triple transgenic maize hybrids where herbicide tolerance and multiple insect resistance traits are integrated as one genomic locus or the combination of transgenes that simultaneously increase synthesis and decrease catabolism of Lys in maize seeds. Recent reports of improvements in gene targeting technology and the construction of meiotically transmissible plant minichromosomes pave the way for introducing more traits with increasing complexity. With such advances, biotechnology is now poised to assemble useful genetic diversity from essentially any source into constructs that concentrate favorable gene action and maximize heritability for a greatly expanded set of traits.

The molecular cloning of QTLs has yielded novel insights about the biology of quantitative traits that were not likely to be discovered from the analysis of gene knockouts or overexpression strategies, in particular the impacts of regulatory variation on phenotypic variation and evolution. Furthermore, molecular markers, genomics, and biotechnology are now applied in an iterative network to exploit genetic diversity for crop improvement. Genomic information enables the discovery of beneficial alleles via QTL mapping and cloning, followed by the use of information learned from the molecular characterization of QTLs to design optimal transgenic strategies for crop improvement.

Molecular Plant Breeding Increases the Efficiency of Selection

Conventional plant breeding that relies only on phenotypic selection has been historically effective. However, for some traits, phenotypic selection has made little progress due to challenges in measuring phenotypes or identifying individuals with the highest breeding value. The effects of environment, genotype by environment interaction, and measurement errors also contribute to observed differences. Evaluation of genotypes in multiple environments with replicated designs allows better estimation of breeding values, but requires additional time and expense. For some traits, it may be necessary to sacrifice the individual to measure phenotypes, or trait expression may depend on variable environmental conditions (e.g. disease pressure) and the stage of development (e.g. grain quality can only be assessed after flowering). Furthermore, plant breeders typically must simultaneously improve a suite of commercially valuable traits, which may limit gains from selection. Just as molecular plant breeding helps to expand genetic diversity, characterize genetic architecture, and modify gene action, its methods can also be applied to increasing the efficiency of selection.

Molecular marker genotypes that are either within genes or tightly linked to QTL influencing traits under selection can be employed as a supplement to phenotypic observations in a selection index. In cases where genetic correlations are high, further efficiencies can be gained by substituting genotypic for phenotypic selection during some selection cycles, which can reduce phenotyping efforts and cycle times by permitting the use of off-season nurseries. Johnson summarized an early example of combining phenotypic data and molecular marker scores to increase selection gains for maize grain yield and resistance to European corn borer. An effective strategy to simultaneously modify multiple traits is the use of selection indices that consider multiple factors in choosing the final improved genotype. Eathington et al. recently reported on results obtained from the use of multiple trait indices and marker-assisted selection for nearly 250 unique corn breeding populations. Use of molecular markers increased breeding efficiency approximately 2-fold relative to phenotypic selection alone, with similar gains also observed in soybean (Glycine max) and sunflower (Helianthus annuus) populations.

Marker-assisted selection can also significantly enhance genetic gain for traits where the phenotype is difficult to evaluate because of its expense or its dependence on specific environmental conditions. Molecular markers may be used to increase the probability of identifying truly superior genotypes, by focusing testing resources on genotypes with the greatest potential (i.e. early elimination of inferior genotypes), by decreasing the number of progeny that must be screened to recover a given level of gain, and by enabling simultaneous improvement for traits that are negatively correlated. Successful examples include resistance to soybean cyst nematode, resistance to cereal diseases, and drought tolerance in maize.

The efficiency of phenotypic selection for some complex traits can be enhanced by including physiological or biochemical phenotypes as secondary traits, if these exhibit strong genetic correlations with the target trait and possess high heritability. Recent advances in functional genomics permit the population-scale profiling of RNA abundance, protein levels and activities, and metabolites that are associated with important traits. In addition to molecular markers that tag DNA sequence variation, such genetical genomics approaches may provide additional secondary phenotypes as selection targets, particularly for traits defined by responses to environmental, developmental, or physiological cues.

Marker-assisted selection also accelerates the deployment of transgenes in commercial cultivars. Typically, this has been achieved through marker-assisted backcrossing. However, for future biotechnology improvements such as tolerance to drought or nutrient limitation, forward breeding may be required to cooptimize transgene expression and genetic background because endogenous genes and environmental factors may have the potential to influence the phenotypes resulting from transgenic modifications. Of course, use of molecular markers could aid forward breeding efforts as well. Alternatively, discovery efforts for additional genes or QTLs that are necessary for dependable trait performance may suggest design of new transgene constructs that stack primary transgenes with known genetic modifiers into second-generation transgenic events.

Increasing Adoption of Molecular Plant Breeding

The adoption of molecular plant breeding approaches has occurred at different rates among crop species and institutions engaged in crop improvement, due to the combined influence of scientific, economic, and sociological factors. Important early scientific barriers included the recalcitrance of cereal crop species to Agrobacterium-mediated transformation and lack of knowledge about genetic control of traits already defined as important breeding targets. Continued research and technology development has largely overcome obstacles for plant transformation of nearly all important crop and horticultural species. Similarly, information gained from genomics research in plant species and other organisms has generated a wealth of information about gene structure and function, as well as large numbers of molecular markers for use in plant breeding. Despite these resources, genetic specificity of robust QTLs remains elusive, unless breeding programs and associated information management systems are restructured to fully integrate knowledge of pedigrees, phenotypes, and marker genotypes that can be leveraged to optimize response to selection. Even with such integration, modifying regulatory functions remains a scientific challenge for molecular breeding because it is difficult to determine the sequence basis for regulatory changes and to predict their phenotypic effects.

Once enabling technologies in biotechnology and genomics become available, economic factors often dictate the degree to which these innovations are integrated into existing plant breeding programs. The expense of gaining governmental regulatory approval for commercial release of transgenic varieties is a significant economic barrier. The costs associated with the development, establishment, and operation of molecular plant breeding are greater than conventional plant breeding practices, requiring significant investments in new research infrastructure and intellectual capacity. Such resources initially existed only in private agribusiness firms and a handful of larger public institutions, further accelerating an ongoing trend for increased industrialization of plant breeding programs among major crops such as corn, soybeans, cotton (Gossypium hirsutum), and wheat. Where there has been adoption in companies, the balance may favor biotechnology over QTLs for improving complex traits, despite greater product development costs, because transgenics can be designed to produce stronger, even dramatic, phenotypic effects, and can often be more rapidly deployed across a broader range of germplasm, resulting in new solutions sooner.

Though molecular breeding is now considered an essential component of current crop improvement efforts for major crops by large companies, the broad applicability of modern molecular approaches to conventional plant breeding remains a source of debate among some practicing plant breeders in the public sector, particularly for minor crops. In addition to the valid scientific and economic factors that have delayed or prevented adoption of molecular approaches in meeting some plant breeding objectives, there are at least three additional reasons contributing to this view. First, molecular plant breeding requires training and expertise in both molecular biology and plant breeding. Educational efforts that delivered such interdisciplinary training were initially established in the early 1990s, but still remain limited to a relatively small group of academic institutions with historic strengths in plant breeding. A second reason for reduced enthusiasm to embrace biotechnology among some plant breeders is the problems with acceptance of transgenic crops among certain governments and groups of consumers, as exemplified by the shelving of wheat varieties with transgenes for resistance to glyphosate herbicides. Finally, excitement about the potential of molecular plant breeding also stimulated shifts in funding at public institutions to enhance intellectual capacity and infrastructure for molecular genetics and genomics research, which ironically often occurred at the expense of conventional plant breeding. This emphasis may have been temporarily necessary to establish the foundations for 21st century plant biology, but there is currently a growing recognition that increased investment in plant breeding capacity and translational research linking molecular methods with breeding objectives is necessary to fully realize the potential of recent advances in biotechnology and genomics.

DOUBLED HAPLOIDY

A doubled haploid (DH) is a genotype formed when haploid cells undergo chromosome doubling. Artificial production of doubled haploids is important in plant breeding.

Haploid cells are produced from pollen or egg cells or from other cells of the gametophyte, then by induced or spontaneous chromosome doubling, a doubled haploid cell is produced, which can be grown into a doubled haploid plant. If the original plant was diploid, the haploid cells are monoploid, and the term doubled monoploid may be used for the doubled haploids. Haploid organisms derived from tetraploids or hexaploids are sometimes called dihaploids (and the doubled dihaploids are, respectively, tetraploid or hexaploid).

Conventional inbreeding procedures take six generations to achieve approximately complete homozygosity, whereas doubled haploidy achieves it in one generation. Dihaploid plants derived from tetraploid crop plants may be important for breeding programs that involve diploid wild relatives of the crops.

Production of Doubled Haploids

Doubled haploids can be produced *in vivo* or *in vitro*. Haploid embryos are produced *in vivo* by parthenogenesis, pseudogamy, or chromosome elimination after wide crossing. The haploid embryo is rescued, cultured, and chromosome-doubling produces doubled haploids. The *in vitro* methods include gynogenesis (ovary and flower culture) and androgenesis (anther and microspore culture). Androgenesis is the preferred method. Another method of producing the haploids is wide crossing. In barley, haploids can be produced by wide crossing with the related species *Hordeum bulbosum*; fertilization is affected, but during the early stages of seed development the *H. bulbosum* chromosomes are eliminated leaving a haploid embryo. In tobacco (*Nicotiana tabacum*), wide crossing with *Nicotiana africana* is widely used. When *N. africana* is used to pollinate *N. tabacum*, 0.25 to 1.42 percent of the progeny survive and can readily be identified as either F1 hybrids or maternal haploids. Although these percentages appear small, the vast yield of tiny seeds and the early death of most seedlings provide significant numbers of viable hybrids and haploids in relatively small soil containers. This method of interspecific pollination serves as a practical way of producing seed-derived haploids of *N. tabacum*, either as an alternative method or complementary method to anther culture.

Genetics of DH Population

In DH method only two types of genotypes occur for a pair of alleles, A and a, with the frequency of ½ AA and ½ aa, while in diploid method three genotypes occur with the frequency of ¼ AA, ½ Aa, ¼ aa. Thus, if AA is desirable genotype, the probability of obtaining this genotype is higher in haploid method than in diploid method. If n loci are segregating, the probability of getting the desirable genotype is (1/2)n by the haploid method and (1/4)n by the diploid method. Hence the efficiency of the haploid method is high when the number of genes concerned is large.

Studies were conducted comparing DH method and other conventional breeding methods and it was concluded that adoption of doubled haploidy does not lead to any bias of genotypes in populations, and random DHs were even found to be compatible to selected line produced by conventional pedigree method.

Applications of DHs Plant Breeding

Mapping Quantitative Trait Loci

Most of the economic traits are controlled by genes with small but cumulative effects. Although the potential of DH populations in quantitative genetics has been understood for some time, it was the advent of molecular marker maps that provided the impetus for their use in identifying loci controlling quantitative traits. As the quantitative trait loci (QTL) effects are small and highly influenced by environmental factors, accurate phenotyping with replicated trials is needed. This is possible with doubled haploidy organisms because of their true breeding nature and because they can conveniently be produced in large numbers. Using DH populations, 130 quantitative traits have been mapped in nine crop species. In total, 56 DH populations were used for QTL detection.

Backcross Breeding

In backcross conversion, genes are introgressed from a donor cultivar or related species into a recipient elite line through repeated backcrossing. A problem in this procedure is being able to identify the lines carrying the trait of interest at each generation. The problem is particularly acute if the trait of interest is recessive, as it will be present only in a heterozygous condition after each backcross. The development of molecular markers provides an easier method of selection based on the genotype (marker) rather than the phenotype. Combined with doubled haploidy it becomes more effective. In marker assisted backcross conversion, a recipient parent is crossed with a donor line and the hybrid (F1) backcrossed to the recipient. The resulting generation (BC1) is backcrossed and the process repeated until the desired genotypes are produced. The combination of doubled haploidy and molecular marker provides the short cut. In the back cross generation one itself a genotype with the character of interest can be selected and converted into homozygous doubled haploid genotype. Chen *et al.* used marker assisted backcross conversion with doubled haploidy of BC1 individuals to select stripe rust resistant lines in barley.

Bulked Segregant Analysis (BSA)

In bulked segregant analysis, a population is screened for a trait of interest and the genotypes at the two extreme ends form two bulks. Then the two bulks are tested for the presence or absence of molecular markers. Since the bulks are supposed to contrast in the alleles that contribute positive and negative effects, any marker polymorphism between the two bulks indicates the linkage between the marker and trait of interest. BSA is dependent on accurate phenotyping and the DH population has particular advantage in that they are true breeding and can be tested repeatedly. DH populations are commonly used in bulked segregant analysis, which is a popular method in marker assisted breeding. This method has been applied mostly to rapeseed and barley.

Genetic Maps

Genetic maps are very important to understand the structure and organization of genomes from which evolution patterns and syntenic relationships between species can be deduced. Genetic maps also provide a framework for the mapping of genes of interest and estimating the magnitude of their effects and aid our understanding of genotype/phenotype associations. DH populations have become standard resources in genetic mapping for species in which DHs are readily available. Doubled haploid populations are ideal for genetic mapping. It is possible to produce a genetic map within two years of the initial cross regardless of the species. Map construction is relatively easy using a DH population derived from a hybrid of two homozygous parents as the expected segregation ratio is simple, *i.e.* 1:1. DH populations have now been used to produce genetic maps of barley, rapeseed, rice, wheat, and pepper. DH populations played a major role in facilitating the generation of the molecular marker maps in eight crop species.

Genetic Studies

Genetic ratios and mutation rates can be read directly from haploid populations. A small doubled haploid (DH) population was used to demonstrate that a dwarfing gene in barley is located chromosome 5H. In another study the segregation of a range of markers has been analyzed in barley.

Genomics

Although QTL analysis has generated a vast amount of information on gene locations and the magnitude of effects on many traits, the identification of the genes involved has remained elusive. This is due to poor resolution of QTL analysis. The solution for this problem would be production of recombinant chromosome substitution line, or stepped aligned recombinant inbred lines. Here, backcrossing is carried out until a desired level of recombination has occurred and genetic markers are used to detect desired recombinant chromosome substitution lines in the target region, which can be fixed by doubled haploidy. In rice, molecular markers have been found to be linked with major genes and QTLs for resistance to rice blast, bacterial blight, and sheath blight in a map produced from DH population.

Elite Crossing

Traditional breeding methods are slow and take 10–15 years for cultivar development. Another disadvantage is inefficiency of selection in early generations because of heterozygosity. These two disadvantages can be over come by DHs, and more elite crosses can be evaluated and selected within less time.

Cultivar Development

Uniformity is a general requirement of cultivated line in most species, which can be easily obtained through DH production. There are various ways in which DHs can be used in cultivar production. The DH lines themselves can be released as cultivars, they may be used as parents in hybrid cultivar production or more indirectly in the creation of breeders lines and in germplasm conservation. Barley has over 100 direct DH cultivars. According to published information there are currently around 300 DH derived cultivars in 12 species worldwide.

The relevance of DHs to plant breeding has increased markedly in recent years owing to the development of protocols for 25 species. Doubled haploidy already plays an important role in hybrid cultivar production of vegetables, and the potential for ornamental production is being vigorously examined. DHs are also being developed in the medicinal herb *Valeriana officinalis* to select lines with high pharmacological activity. Another interesting development is that fertile homozygous DH lines can be produced in species that have self-incompatibility systems.

Advantages of DHs

The ability to produce homozygous lines after a single round recombination saves a lot of time for the plant breeders. Studies conclude that random DH's are comparable to the selected lines in pedigree inbreeding. The other advantages include development of large number of homozygous lines, efficient genetic analysis and development of markers for useful traits in much less time. More specific benefits include the possibility of seed propagation as an alternative to vegetative multiplication in ornamentals, and in species such as trees in which long life cycles and inbreeding depression preclude traditional breeding methods, doubled haploidy provides new alternatives.

Disadvantages of DHs

The main disadvantage with the DH population is that selection cannot be imposed on the population. But in conventional breeding selection can be practised for several generations: thereby desirable characters can be improved in the population.

In haploids produced from anther culture, it is observed that some plants are aneuploids and some are mixed haploid-diploid types. Another disadvantage associated with the double haploidy is the cost involved in establishing tissue culture and growth facilities. The over-usage of doubled haploidy may reduce genetic variation in breeding germplasm. Hence one has to take several factors into consideration before deploying doubled haploidy in breeding programmes.

BREEDING FOR DROUGHT STRESS TOLERANCE

Breeding for drought resistance is the process of breeding plants with the goal of reducing the impact of dehydration on plant growth.

Dehydration Stress in Crop Plants

In nature or crop fields, water is often the most limiting factor for plant growth. If plants do not receive adequate rainfall or irrigation, the resulting dehydration stress can reduce growth more than all other environmental stresses combined.

Drought can be defined as the absence of rainfall or irrigation for a period of time sufficient to deplete soil moisture and cause dehydration in plant tissues. Dehydration stress results when water loss from the plant exceeds the ability of the plant's roots to absorb water and when the plant's water content is reduced enough to interfere with normal plant processes.

Dehydration Stress is a Global Phenomenon

About 15 million km2 of the land surface is covered by crop-land, and about 16% of this area is equipped for irrigation. Thus, in many parts of the world, including the United States, plants may frequently encounter dehydration stress. Rainfall is very seasonal and periodic drought occurs regularly. The effect of drought is more prominent in sandy soils with low water holding capacity. On such soils some plants may experience dehydration stress after only a few days without water.

During the 20th century, the rate of increase in `blue' water withdrawal (from rivers, lakes, and aquifers) for irrigation and other purposes was higher than the growth rate of the world population. Country-wise maps of irrigated areas are available.

Dehydration Stress and Future Challenges to Crop Production

Soil moisture deficit is a significant challenge to the future of crop production. Severe drought in parts of the U.S., Australia, and Africa in recent years drastically reduced crop yields and disrupted regional economies. Even in average years, however, many agricultural regions, including the U.S. Great Plains, suffer from chronic soil moisture deficits. Cereal crops typically attain only about 25% of their potential yield due to the effects of environmental stress, with dehydration stress the most important cause. Two major trends will likely increase the frequency and severity of soil moisture deficits.

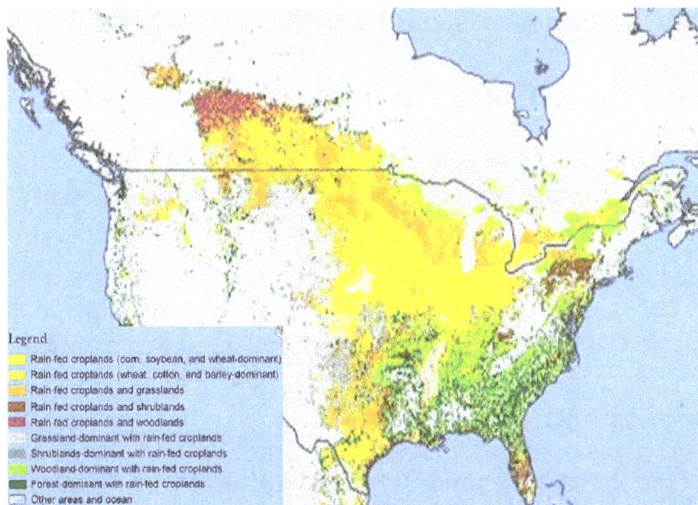

Rain fed areas of USA.

Climate Change

Higher temperatures are likely to increase crop water use due to increased transpiration. A warmer atmosphere will also speed up melting of mountain snow pack, resulting in less water available for irrigation. More extreme weather patterns will increase the frequency of drought in some regions.

Competing uses for Limited Water Supplies

Increased demand from municipal and industrial users will further reduce the amount of water available for irrigated crops.

Although changes in tillage and irrigation practices can improve production by conserving water, enhancing the genetic tolerance of crops to drought stress is considered an essential strategy for addressing moisture deficits.

Dehydration Stress affects Plant Physiology

A plant responds to a lack of water by halting growth and reducing photosynthesis and other plant processes in order to reduce water use. As water loss progresses, leaves of some species may appear to change colour — usually to blue-green. Foliage begins to wilt and, if the plant is not irrigated, leaves will fall off and the plant will eventually die. Soil moisture deficit lowers the water potential of a plant's root and, upon extended exposure, abscisic acid is accumulated and eventually stomatal closure occurs. This reduces a plant's leaf relative water content. The time required for dehydration stress to occur depends on the water-holding capacity of the soil, environmental conditions, stage of plant growth, and plant species. Plants growing in sandy soils with low water-holding capacity are more susceptible to dehydration stress than plants growing in clay soils. A limited root system will accelerate the rate at which dehydration stress develops. A plant's root system may be limited by the presence of competing root systems from neighbouring plants, by site conditions such as compacted soils or high water tables, or by container size (if growing in a container). A plant with a large mass of leaves in relation to the root system is prone to drought stress because the leaves may lose water faster than the roots can supply it. Newly planted plants and poorly established plants may be especially susceptible to dehydration stress because of the limited root system or the large mass of stems and leaves in comparison to roots.

Dehydration Stress Interaction with other Stress Factors

Aside from the moisture content of the soil, environmental conditions of high light intensity, high temperature, low relative humidity and high wind speed will significantly increase plant water loss. The prior environment of a plant also can influence the development of dehydration stress. A plant that has been exposed to dehydration stress (hardened) previously and has recovered may become more drought resistant. Also, a plant that was well-watered prior to being water-limited will usually survive a period of drought better than a continuously dehydration-stressed plant.

Mechanisms of Drought Resistance

The degree of resistance to drought depends upon individual crops. Generally three strategies can help a crop to mitigate the effect of dehydration stress.

Avoidance

If the plant shows dehydration avoidance, the environmental factor is excluded from the plant tissues by reducing water loss ("water savers", e.g. by thick leaf epicuticular wax, leaf rolling, leaf posture) or maintaining water uptake ("water spenders", e.g. by growing deeper roots). Dehydration avoidance is desirable in modern agriculture, where drought resistance requires the maintenance of economically viable plant production under dehydration stress. The role of dehydration avoidance is maintaining water supply and sustaining leaf hydration and turgidity with the purpose of maintaining stomatal opening and transpiration as long as possible under water deficit. This is essential for leaf gas exchange, photosynthesis and plant production through carbon assimilation.

Tolerance

If the plant shows dehydration tolerance, the environmental factor enters the plant tissues but the tissues survive, by e.g. maintaining turgor and osmotic adjustment.

Escape

Dehydration escape involves e.g. early maturing or seed dormancy, where the plant uses previous optimal conditions to develop vigor. Dehydration Recovery refers to some plant species being able to recuperate after brief drought periods.

A proper timing of life-cycle, resulting in the completion of the most sensitive developmental stages while water is abundant, is considered to be a dehydration escape strategy. Avoiding dehydration stress with a root system capable of extracting water from deep soil layers, or by reducing evapotranspiration without affecting yields, is considered as dehydration avoidance. Mechanisms such as osmotic adjustment (OA) whereby a plant maintains cell turgor pressure under reduced soil water potential are categorised as dehydration tolerance mechanisms. Dehydration avoidance mechanisms can be expressed even in the absence of stress and are then considered constitutive. Dehydration tolerance mechanisms are the result of a response triggered by dehydration stress itself and are therefore considered adaptive. When the stress is terminal and predictable, dehydration escape through the use of shorter duration varieties is often the preferable method of improving yield potential. Dehydration avoidance and tolerance mechanisms are required in situations where the timing of drought is mostly unpredictable.

Drought resistance mechanisms are genetically controlled and genes or QTL responsible for drought resistance have been discovered in several crops which opens avenue for molecular breeding for drought resistance.

Drought Resistance Traits

Resistance to drought is a quantitative trait, with a complex phenotype, often confounded by plant phenology. Breeding for drought resistance is further complicated since several types of abiotic stress, such as high temperatures, high irradiance, and nutrient toxicities or deficiencies can challenge crop plants simultaneously.

Osmotic Adjustment

When a plant is exposed to water deficit, it may accumulate a variety of osmotically active compounds such as amino acids and sugars, resulting in a lowering of the osmotic potential. Examples of amino acids that may be up-regulated are proline and glycine betaine. This is termed osmotic adjustment and enables the plant to take up water, maintain turgor and survive longer.

Cell Membrane Stability

The ability to survive dehydration is influenced by a cell's ability to survive at reduced water content. This can be considered complementary to OA because both traits will help maintain leaf growth (or prevent leaf death) during water deficit. Crop varieties differ in dehydration tolerance and an important factor for such differences is the capacity of the cell membrane to prevent electrolyte leakage at decreasing water content, or "cell membrane stability (CMS)". The maintenance of membrane function is assumed to mean that cell activity is also maintained. Measurements of CMS have been used in different crops and are known to be correlated with yields under high temperature and possibly under dehydration stress.

Epicuticular Wax

In sorghum (Sorghum bicolor L. Moench), drought resistance is a trait that is highly correlated with the thickness of the epicuticular wax layer. Experiments have demonstrated that rice varieties with a thick cuticle layer retain their leaf turgor for longer periods of time after the onset of a water-stress.

Partitioning and Stem Reserve Mobilisation

As photosynthesis becomes inhibited by dehydration, the grain filling process becomes increasingly reliant on stem reserve utilisation. Numerous studies have reported that stem reserve mobilisation capacity is related to yield under dehydration stress in wheat. In rice, a few studies also indicated that this mechanism maintains grain yield under dehydration stress at the grain filling stage. This dehydration tolerance mechanism is stimulated by a decrease in gibberellic acid concentration and an increase in abscisic acid concentration.

Manupulation and Stability of Flowering Processes

Seedling Drought Resistance Traits

For emergence from deep sowing (to exploit dry upper soil), this is practised to help seedlings reach the receding moisture profile, and to avoid high soil surface temperatures which inhibit germination. Screening at these stage provides practical advantages, specially when managing large amount of germ-plasms.

Drought Resistant Ideotype

Usually ideotypes are developed to create an ideal plant variety. The following traits constitutes ideotype of wheat by CIMMYT:

- Large seed size. Helps emergence, early ground cover, and initial biomass.

- Long coleoptiles for emergence from deep sowing.

- Early ground cover.

Thinner, wider leaves (i.e., with a relatively low specific leaf weight) and a more prostrate growth habit help to increase ground cover, thus conserving soil moisture and potentially increasing radiation use efficiency:

- High pre-anthesis biomass.

- Good capacity for stem reserves and remobilisation.

- High spike photosynthetic capacity.

- High RLWC/Gs/CTD during grain filling to indicate ability to extract water.

- Osmotic adjustment.

- Accumulation of ABA.

The benefit of ABA accumulation under dehydration has been demonstrated. It appears to pre-adapt plants to stress by reducing stomatal conductance, rates of cell division, organ size, and increasing development rate. However, high ABA can also result in sterility problems since high ABA levels may abort developing florets:

- Heat Tolerance: The contribution of heat tolerance to performance under dehydration stress needs to be quantified, but it is relatively easy to screen for.

- Leaf anatomy: Waxiness, pubescence, rolling, thickness, posture. These traits decrease radiation load to the leaf surface. Benefits include a lower evapotranspiration rate and reduced risk of irreversible photo-inhibition. However, they may also be associated with reduce radiation use efficiency, which would reduce yield under more favourable conditions.

- High tiller survival: Comparison of old and new varieties have shown that under dehydration older varieties over-produce tillers many of which fail to set grain while modern drought resistant lines produce fewer tillers most of which survive.

- Stay-green: The trait may indicate the presence of drought resistance mechanisms, but probably does not contribute to yield per se if there is no water left in the soil profile by the end of the cycle to support leaf gas exchange. It may be detrimental if it indicates lack of ability to remobilise stem reserves. However, research in sorghum has indicated that Stay-green is associated with higher leaf chlorophyll content at all stages of development and both were associated with improved yield and transpiration efficiency under dehydration.

Combination Phenomics

The concept of combination phenomics comes from the idea that two or more plant stresses have common physiological effects or common traits - which are an indicator of overall plant health. As both biotic and abiotic stresses can result in similar physiological consequence, drought resistant plants can be separated from sensitive plants. Some imaging or infrared measuring techniques can help to speed the process for breeding process. For example, spot blotch intensity and canopy temperature depression can be monitored with canopy temperature depression.

Molecular Breeding for Drought Resistance

Recent research breakthroughs in biotechnology have revived interest in targeted drought resistance breeding and use of new genomics tools to enhance crop water productivity. Marker-assisted breeding is revolutionising the improvement of temperate field crops and will have similar impacts on breeding of tropical crops. Other molecular breeding tool include development of genetically modified crops that can tolerate plant stress. As a complement to the recent rapid progress in genomics, a better understanding of physiological mechanisms of dehydration response will also contribute to the progress of genetic enhancement of crop drought resistance. It is now well accepted that the complexity of the dehydration syndrome can only be tackled with a holistic approach that integrates physiological dissection of crop dehydration avoidance and - tolerance traits using molecular genetic tools such as MAS, micro-arrays and transgenic crops, with agronomic practices that lead to better conservation and utilisation of soil moisture, and better matching of crop genotypes with the environment. MAS has been implemented in rice varieties to assess the drought tolerance and to develop new abiotic stress-tolerant varieties.

BREEDING FOR HEAT STRESS TOLERANCE

Plant breeding is process of development of new cultivars. Plant breeding involves development of varieties for different environmental conditions – some of them are not favorable. Among them, heat stress is one of such factor that reduces the production and quality significantly. So breeding against heat is a very important criterion for breeding for current as well as future environments produced by global climate change (e.g. global warming).

In Plants

Heat stress due to increased temperature is a very important problem globally. Occasional or prolonged high temperatures cause different morpho-anatomical, physiological and biochemical changes in plants. The ultimate effect is on plant growth as well as development and reduced yield and quality. Breeding for heat stress tolerance can be mitigated by breeding plant varieties that have improved levels of thermo-tolerance using different conventional or advanced genetic tools. Marker assisted selection techniques for breeding are highly useful. Recently 41 polymorphic SSR markers has been identified between a heat tolerant rice variety 'N22' and heat susceptible-high yielding variety 'Uma' for the development of new 'high yielding-heat tolerant' rice varieties.

Heat Stress Tolerance

Heat stress is defined as increased temperature level sufficient to cause irreversible damage to plant growth and development. Generally a temperature rise, above usually 10 to 15 °C above ambient, can be considered heat shock or heat stress. Heat tolerance is broadly defined as the ability of the plant tolerate heat – means that grow and produce economic yield under high temperatures.

Significance of Global Warming

Heat stress is a serious threat to crop production globally. Global warming is particularly consequence of increased level of green house gases such as CO_2, methane, chlorofluorocarbons and nitrous oxides. The Intergovernmental Panel on Climatic Change (IPCC) has predicted a rise of 0.3 °C per decade reaching to approximately 1 and 3 °C above the present value by 2025 and 2100 AD, respectively.

Physiological Consequence of Heat Stress

At very high temperatures cause severe cellular injury and cell death may occur within short time, thus leading to a catastrophic collapse of cellular organization. However, under moderately high temperatures, the injury can only occur after longer exposure to such a temperature however the plant efficiency can be severely affected. High temperature directly affect injuries such as protein denaturation and aggregation, and increased fluidity of membrane lipids. Other indirect or slower heat injuries involve inactivation of enzymes in chloroplast and mitochondria, protein degradation, inhibition of protein synthesis, and loss of membrane integrity. Heat stress associated injuries ultimately lead to starvation, inhibition of growth, reduced ion flux, production of toxic compounds and production of reactive oxygen species (ROS). Immediately after exposure to high temperature stress-related proteins are expressed as stress defense strategy of the cell. Expression of heat shock proteins (HSPs), protein with 10 to 200 kDa, is supposed to be involved in signal transduction during heat stress. In many species it has been demonstrated that HSPs results in improved physiological phenomena such as photosynthesis, assimilate partitioning, water and nutrient use efficiency, and membrane stability.

Studies have found tremendous variation within and between species, thus this will help to breed heat tolerance for future environment. Some of attempts to develop heat-tolerant genotypes are successful.

Traits Associated with Heat Stress Tolerance

Different physiological mechanisms may contribute to heat tolerance in the field—for example, heat tolerant metabolism as indicated by higher photosynthetic rates, stay-green, and membrane thermo-stability, or heat avoidance as indicated by canopy temperature depression. Several physiological and morphological traits have been evaluated for heat tolerance - Canopy temperature, leaf chlorophyll, stay green, leaf conductance, spike number, biomass, and flowering date.

(a) Canopy temperature depression (CTD): CTD has shown clear association with yield in warm environments shows it association with heat stress tolerance. CTD shows high genetic correlation with yield and high values of proportion of direct response to selection indicating that the trait is heritable and therefore amenable to early generation selection. Since an integrated CTD value can be measured almost instantaneously on scores of plants in a small breeding plot (thus reducing error normally associated with traits measured on individual plants), work has been conducted to evaluate its potential as an indirect selection criterion for genetic gains in yield. CTD is affected by many physiological factors, which makes it a powerful.

(b) Stomatal conductance: Canopy temperature depression is highly suitable for selecting physiologically superior lines in warm, low relative humidity environments where high evaporative demand leads to leaf cooling of up to 10 °C below ambient temperatures. This permits differences among genotypes to be detected relatively easily using infrared thermometry. However, such differences cannot be detected in high relative humidity environments because the effect of evaporative cooling of leaves is negligible. Nonetheless, leaves maintain their stomata open to permit the uptake of CO_2, and differences in the rate of CO_2 fixation may lead to differences in leaf conductance that can be measured using a porometer. Porometry can be used to screen individual plants. The heritability of stomatal conductance is reasonably high, with reported values typically in the range of 0.5 to 0.8. Plants can be assessed for leaf conductance using a viscous flow porometer that is available on the market (Thermoline and CSIRO, Australia). This instrument can give a relative measure of stomatal conductance in a few seconds, making it possible to identify physiologically superior genotypes from within bulks.

(c) Membrane thermostability: Although resistance to high temperatures involves several complex tolerance and avoidance mechanisms, the membrane is thought to be a site of primary physiological injury by heat, and measurement of solute leakage from tissue can be used to estimate damage to membranes. Since membrane thermostability is reasonably heritable and shows high genetic correlation with yield.

(d) Chlorophyll fluorescence: Chlorophyll fluorescence, an indication of the fate of excitation energy in the photosynthetic apparatus, has been used indicator for heat stress tolerance.

(e) Chlorophyll content and stay green: Chlorophyll content and stay green traits have been found to be associated with heat stress tolerance. Xu et al. identified three QTLs for chlorophyll content (Chl1, Chl2, and Chl3) (coincided with three stay-green QTL regions (Stg1, Stg2, and Stg3)) were identified in Sorghum. The Stg1 and Stg2 regions also contain the genes for key photosynthetic enzymes, heat shock proteins, and an abscisic acid (ABA) responsive gene.

(f) Photosynthesis: Decliend photosynthesis is suggested as measure of heat stress sensitivity in plants.

(g) Stem reserve remobilization.

Combination Breeding and Physiological Breeding

The physiological-trait-based breeding approach has merit over breeding for yield *per se* because it increases the probability of crosses resulting in additive gene action. The concept of combination phenomics comes from the idea that two or more stress have common physiological effect or common traits - which is an indicator of overall plant health. Similar analogy in human medical terms is high blood pressure or high body temperature or high white blood cells in body is an indicator of health problems and thus we can select healthy people from unhealthy using such a measure. As both abiotic and abiotic stresses can result in similar physiological consequence, tolerant plant can be separated from sensitive plants. Some imaging or infrared measuring techniques can help to speed the process for breeding process. For example, spot blotch intensity and canopy temperature depression can be monitored with canopy temperature depression.

MARKER-ASSISTED BACKCROSSING

Marker-assisted or marker-based backcrossing (MABC) is regarded as the simplest form of marker-assisted selection, and at the present it is the most widely and successfully used method in practical molecular breeding. MABC aims to transfer one or a few genes/QTLs of interest from one genetic source (serving as the donor parent and maybe inferior agronomically or not good enough in comprehensive performance in many cases) into a superior cultivar or elite breeding line (serving as the recurrent parent) to improve the targeted trait. Unlike traditional backcrossing, MABC is based on the alleles of markers associated with or linked to gene(s)/QTL(s) of interest instead of phenotypic performance of target trait. The general procedure of MABC is as follow, regardless of dominant or recessive nature of the target trait in inheritance:

- Select parents and make the cross, one parent is superior in comprehensive performance and serves as recurrent parent (RP), and the other one used as donor parent (DP) should possess the desired trait and the DNA markers allele(s) associated with or linked to the gene for the trait.

- Plant F_1 population and detect the presence of the marker allele(s) at early stages of growth to eliminate false hybrids, and cross the true F_1 plants back to the RP.

- Plant BCF_1 population, screen individuals for the marker(s) at early growth stages, and cross the individuals carrying the desired marker allele(s) (in heterozygous status) back to the RP. Repeat this step in subsequent seasons for two to four generations, depending upon the practical requirements and operation situations as discussed below.

- Plant the final backcrossing population (e.g. BC_4F_1), and screen individual plants with the marker(s) for the target trait and discard the individuals carrying homozygous markers alleles from the RP. Have the individuals with required marker allele(s) selfed and harvest them.

- Plant the progenies of backcrossing-selfing (e.g. BC_4F_2), detect the markers and harvest individuals carrying homozygous DP marker allele(s) of target trait for further evaluation and release.

Theoretically, the proportion of the RP genome after n generations of backcrossing is given by $1 - (1/2)^{n+1}$ for a single locus and $[1 - (1/2)^{n+1}]k$ for k loci, respectively, for a population large enough in size (or with adequate individuals) and no selection being made during backcrossing (i.e. "blind" backcrossing only). The percentage of the RP genome is the average of the population, with some individuals possessing more of the RP genome than others. To fully recover the genome of the RP, 6-8 generations of backcrossing is needed typically in case no selection is made for the RP. However, this process is usually slower than expected for the target gene-carrier chromosome, i.e. linkage drag, especially in case a linkage exists between the target gene and other undesirable traits. On the other hand, the process of introgression of QTLs/genes and recovery of the RP genome may be accelerated by selection using markers flanking QTLs and evenly spaced markers from other chromosomes (i.e. unlinked to QTLs) of the RP or selection for the performance of the RP conducted simultaneously. For MABC program, therefore, there are two types of selection recognized: Foreground selection and background selection.

In foreground selection, the selection is made only for the marker allele(s) of donor parent at the target locus to maintain the target locus in heterozygous state until the final backcrossing is completed. Then the selected plants are selfed and the progeny plants with homozygous DP allele(s) of selected markers are harvested for further evaluation and release. As described above, this is the general procedure of MABC. The effectiveness of foreground selection depends on the number of genes/loci involved in the selection, the marker-gene/QTL association or linkage distance and the undesirable linkage to the target gene/QTL.

In background selection, the selection is made for the marker alleles of recurrent parent in all genomic regions of desirable traits except the target locus, or selection against the undesirable genome of donor parent. The objective is to hasten the restoration of the RP genome and eliminate undesirable genes introduced from the DP. The progress in recovery of the RP genome depends on the number of markers used in background selection. The more markers evenly located on all the chromosomes are selected for the RP alleles, the faster recovery of the RP genome will be achieved but larger population size and more genotyping will be required as well. In addition, the linkage drag also can be efficiently addressed by background selection using DNA markers, although it is difficult to overcome in a traditional backcrossing program.

Foreground selection and background selection are two respective aspects of MABC with different foci of selection. In practice, however, both foreground and background selection are usually conducted in the same program, either simultaneously or successively. In many cases, they can be performed alternatively even in the same generation. The individuals that have the desired marker alleles for target trait are selected first (foreground selection). Then the selected individuals are screened for other marker alleles again for the RP genome (background selection). It is understandable to do so because selection of the target gene/QTL is the essential and only critical point for backcrossing program, and the individuals that do not have the allele of target gene will be discarded and thus it is not necessary to genotype them for other traits.

The efficiency of MABC depends upon several factors, such as the population size for each generation of backcrossing, marker-gene association or the distance of markers from the target locus, number of markers used for target trait and RP background, and undesirable linkage drag. Based on simulations of 1000 replicates, Hospital presented the expected results of a typical MABC program, in which heterozygotes were selected at the target locus in each generation, and RP alleles were selected for two flanking markers on target chromosome each located 2 cM apart from the target locus and for three markers on non-target chromosomes. As shown in table, a faster recovery of the RP genome could be achieved by MABC with combined foreground and background selection, compared to traditional backcrossing. Therefore, using markers can lead to considerable time savings compared to conventional backcrossing.

Backcross generation	Number of individuals	% Homozygosity of recurrent parent alleles at selected markers		% Recurrent parent genome	
		Chromosome with target locus	All other chromosomes	Marker-assisted backcross	Conventional backcross
BC_1	70	38.4	60.6	79.0	75.0
BC_2	100	73.6	87.4	92.2	87.5
BC_3	150	93.0	98.8	98.0	93.7
BC_4	300	100.0	100.0	99.0	96.9

Expected results of a MABC program with combined foreground and background selection used. In a MABC program, the population to be analyzed should contain at least one genotype that has all favorable alleles for a particular QTL. Later, the number of QTLs may be increased progressively, but not beyond six QTLs in most cases because of prohibitive difficulty in handling all QTLs.

In addition, the more QTLs/genes are transferred, the larger the proportion of unwanted genes would be due to linkage drag. In general, most of the unwanted genes are located on non-target chromosomes in early BC generations, and are rapidly removed in subsequent BC generations. On the contrary, the quantity of DP genes on the target chromosome decreases much more slowly, and even after generation BC_6 many of the unwanted donor genes are still located on the target chromosome in segregating state. Given a total genome length is 3000 cM, 1% donor DNA fragments after six backcrosses represents a 30 cM chromosomal segment or region, which may host many unwanted genes, especially if the DP is a wild genetic resource. Young and Tanksley genotyped a collection of tomato varieties in which the resistance gene was previously transferred at the Tm-2 locus with RLFP markers. Their data indicated that the size of chromosomal segment retained around the Tm-2 locus during backcross breeding was very variable, with one line exhibiting a donor segment of 50 cM after 11 backcrosses and other one possessing 36 cM donor segment after 21 backcrosses. This clearly demonstrates the need for background selection.

Linkage drag can be reduced by performing background selection. Typically, two markers flanking the target gene are used, and the individuals (or double recombinants) that are heterozygous at the target locus and homozygous for the recipient (RP) alleles at both flanking markers are selected. Use of closer flanking markers leads to more effective and faster reduction of linkage drag compared to distant markers. However, less distance between two flanking markers implies less probability of double recombination, and thus larger populations and more genotyping are needed. In order to optimize genotyping effort (i.e. the cost of the program), therefore, it is important to determine the minimal population sizes necessary to ensure the desired genotypes can be obtained. Hospital and Decoux developed a statistical software for determining the minimum population size required in BC program to identify at least one individual that is double-recombinant with heterozygosity at target locus and homozygosity for recurrent parent alleles at flanking marker loci. In addition, for closely-linked flanking markers, it is unlikely to obtain double recombinant genotypes through only one generation of backcrossing. Therefore, additional backcrossing should be conducted. For instance, in one BC generation (e.g. BC1) single recombination on one side of the target gene is selection, and single recombination on the other side may be selected in another BC generation (e.g. BC_2). In this way, individuals with desired RP alleles at two flanking markers and donor allele at target locus can be finally obtained.

To accelerate the recovery of RP genome on non-target chromosomes, scientists suggested using markers in backcrossing and discussed how many makers should be used. In background selection, the approaches involve selecting individuals that are of homozygous recipient type at a collection of markers located on non-carrier chromosomes. From a point of both effectiveness and efficiency, it is important to determine an appropriate number of markers to be used. More markers do not necessarily mean better benefits in practice. Generally, several markers are involved and MABC should be performed over two or more generations. It is unlikely that the selection objective can be realized in a single BC generation.

Dense marker coverage of non-target chromosomes is not mandatory to increase the overall proportion of recurrent parent genome, unless fine-mapping of specific chromosome regions is highly important. An appropriate number of markers and optimal position on chromosomes are important. Computer simulation suggested that for a chromosome of 100 cM, two to four markers are sufficient, and selection based on markers would be most efficient if the markers are optimally positioned along the chromosomes. In practice, at least two or three markers per chromosome are needed, and every chromosome should be involved. In such a MABC scheme, three to four generations of backcrossing is generally enough to achieve more than 99% of the recurrent parent genome. With respect to the time necessary to release new varieties, the gain due to background selection can be economically valuable. In addition, background selection is more efficient in late BC generations than in early BC generations. For example, if a BC breeding scheme is conducted over three successive BC generations and yet the preference is to genotype individuals only once, then it is more efficient to genotype and select the individuals in BC_3 generation rather than in the BC_1 generation.

Application of MABC

Success in integrating MABC as a breeding approach lies in identifying situations in which markers offer noticeable advantages over conventional backcrossing or valuable complements to conventional breeding effort. MABC is essential and advantageous when:

- Phenotyping is difficult and/or expensive or impossible;

- Heritability of the target trait is low;

- The trait is expressed in late stages of plant development and growth, such as flowers, fruits, seeds, etc.;

- The traits are controlled by genes that require special conditions to express;

- The traits are controlled by recessive genes; and

- Gene pyramiding is needed for one or more traits.

Among the molecular breeding methods, MABC has been most widely and successfully used in plant breeding up to date. It has been applied to different types of traits (e.g. disease/pest resistance, drought tolerance and quality) in many species, e.g. rice, wheat, maize, barley, pear millet, soybean, tomato, etc. In maize, for example, Bacillus thuringiens is a bacterium that produces insecticidal toxins, which can kill corn borer larvae when they ingest the toxins in corn cells. The integration of the Bt transgene into various corn genetic backgrounds has been achieved by using MABC. Aroma in rice is controlled by a recessive gene which is due to an eight base-pair deletion and three single nucleotide polymorphism in a gene that codes for betaine aldehyde dehydrgenase 2. This discovery allows identification of the aromatic and non-aromatic rice varieties and discriminates homozygous recessive and dominant as well as heterozygous individuals in segregating population for the trait. MABC has been used to select for aroma in rice. High lysine opaque2 gene in corn was incorporated using MABC. However, the rate of success decreases when large numbers of QTLs are targeted for introgression. Sebolt et al. used MABC for two QTL for seed protein content in soybeans. However, only one QTL was confirmed in $BC_3F_{4:5}$. When that QTL was introduced in three different genetic backgrounds, it had no effect in one background. In tomato, Tanksley and Nelson proposed a MABC strategy, called advanced backcross-QTL (AB-QTL), to transfer resistance genes from wild relative/unadapted genotype into elite germplasm. The strategy has proven effective for various agronomically important traits in tomato, including fruit quality and black mold resistance. In addition, AB-QTL has been used in other crop species, such as rice, barley, wheat, maize, cotton and soybean, collectively demonstrating that this strategy is effective in transferring favorable alleles from the wild/unadapted germplasm to elite germplasm.

In barley, a marker linked (0.7 cM) to the Yd2 gene for resistance to barley yellow dwarf virus was successfully used to select for resistance in a backcrossing scheme. Compared to lines without the marker, the BC2F2-derived lines carrying the linked marker had lighter leaf symptoms and higher yield when infected by the virus. In maize, marker-facilitated backcrossing was also successfully employed to improve complex traits such as grain yield. Using MABC, six chromosomal segments each in two elite lines, Tx303 and Oh43, were transferred into two widely used inbred lines, B73 and Mo17, through three generations of backcrossing followed by two selfing generations. Then the enhanced lines with better performance were selected based on initial evaluations of testcross hybrids. The single-cross hybrids of enhanced B73 x enhanced Mo17 out-yielded the check hybrids by 12-15%. Zhao et al. reported that a major quantitative trait locus (named qHSR1) for resistance to head smut in maize was successfully integrated into ten high-yielding inbred lines (susceptible to head smut). Each of the ten high-yielding lines was crossed with a donor parent Ji 1037 that contains qHSR1 and is completely resistant to head smut, followed by five generations of backcrossing to the respective recurrent parents. In BC1 through BC3 only phenotypic selection was conducted to identify highly resistant individuals after artificial inoculation. In BC4 phenotypic selection, foreground selection and recombinant selection were conducted to screen for resistant individuals with the shortest qHSR1 donor regions. In BC5, phenotypic selection, foreground selection and background selection were performed to identify resistant individuals with the highest proportion of the recurrent parent genome, followed by one generation of self-pollination to obtain homozygous genotypes at the qHSR1 locus. The ten improved inbred lines all showed substantial resistance to head smut, and the hybrids derived from these lines also showed a significant increase in the resistance. Semagn et al. provided a detail review on the progress and prospects of MABC in crop breeding.

Currently, a cooperative marker-based backcrossing project for high-oleic acid in soybean has been initiated among multiple U.S. land-grant universities and USDA-ARS. Backcrossing and selection will be performed using the markers tightly linked to the high-oleic genes/loci. Hopefully, the high-oleic (80% or higher) traits will be successfully transferred from mutant lines or derived lines into other locally superior cultivars/lines, or combined with other unique traits like low linolenic acid.

Marker-assisted Gene Pyramiding and Marker-assisted Recurrent Selection

Marker-assisted gene pyramiding (MAGP) is one of the most important applications of DNA markers to plant breeding. Gene pyramiding has been proposed and applied to enhance resistance to disease and insects by selecting for two or more than two genes at a time. For example in rice such pyramids have been developed against bacterial blight and blast. Castro et al. reported a success in pyramiding qualitative gene and QTLs for resistance to stripe rust in barley. The advantage of using markers in this case allows selecting for QTL-allele-linked markers that have the same phenotypic effect. To enhance or improve a quantitatively inherited trait in plant breeding, pyramiding of multiple genes or QTLs is recommended as a potential strategy. The cumulative effects of multiple-QTL pyramiding have been proven in crop species like wheat, barley and soybean. Pyramiding of multiple genes/QTLs may be achieved through different approaches: multiple-parent crossing or complex crossing, backcrossing, and recurrent selection. A suitable breeding scheme for MAGP depends on the number of genes/QTLs required for improvement of traits, the number of parents that contain the required genes/QTLs, the heritability of traits of interest, and other factors (e.g. marker-gene association, expected duration to complete the plan and relative cost). Assuming three or four desired genes/QTLs exist separately in three or four lines, pyramiding of them can be realized by three-way, four-way or double crossing. They may also be integrated by convergent backcrossing or stepwise backcrossing. However, if there are more than four genes/QTLs to be pyramided, complex or multiple crossing and/or recurrent selection may be often preferred.

For MABC-based gene pyramiding, in general, there may be three strategies or breeding schemes: stepwise, simultaneous/synchronized and convergent backcrossing or transfer. Supposing one cultivar W is superior in comprehensive performance but lack of a trait of interest, and four different genes/QTLs contributing to the trait have been identified in four germplasm lines (e.g. P1, P2, P3 and P4). Three MABC schemes for pyramiding the genes/QTLs can be described as follows:

Scheme 1: Stepwise Backcrossing;

Scheme 2: Simultaneous/Synchronized Backcrossing;

Scheme 3: Convergent Backcrossing.

In the stepwise backcrossing, four target genes/QTLs are transferred into the recurrent parent W in order. In one step of backcrossing, one gene/QTL is targeted and selected, followed by next step of backcrossing for another gene/QTL, until all target genes/QTLs have been introgressed into the RP. The advantage is that gene pyramiding is more precise and easier to implement as it involves only one gene/QTL at one time and thus the population size and genotyping amount will be small. The improved recurrent parent may be released before the final step as long as the integrated genes/QTLs (e.g. two or three) meet the requirement at that time. The disadvantage is that it takes a longer time to complete. In the simultaneous or synchronized backcrossing, the recurrent parent W is first crossed to each of four donor parents to produce four single-cross F1s. Two of the four single-cross F1s are crossed with each other to produce two double-cross F1s, and these two double-cross F1s are crossed again to produce a hybrid integrating all four target genes/QTLs in heterozygous state. The hybrid and/or progeny with heterozygous markers for all four target genes/QTLs is subsequently crossed back to the RP W until a satisfactory recovery of the RP genome, and finalized by one generation of selfing. The advantage of this method is that it takes the shortest time to complete. However, in the backcrossing all target genes/QTLs are involved at the same time and thus it requires a large population and more genotyping. Convergent backcrossing is a strategy combining the advantages of stepwise and synchronized backcrossing. First the four target gene/QTLs are transferred separately from the donors into the recurrent parent W by single crossing followed by backcrossing based on markers linked to the target genes/QTLs, to produce four improved lines (WAA, WBB, WCC, and WDD). Two of the improved lines are crossed with each other and the two hybrids are then intercrossed to integrate all four genes/QTLs together and develop the final improved line with all four genes/QTLs pyramided (i.g. WAA BBC CDD). Relatively speaking, convergent backcrossing is more acceptable because in this scheme not only is time reduced (compared to stepwise transfer) but gene fixation and/or pyramiding is also more easily assured (compared to simultaneous transfer).

Theoretical issues and efficiency of MABC for gene pyramiding have been investigated through computer simulations. Practical application of MABC to gene pyramiding has been reported in many crops, including rice, wheat, barley, cotton, soybean, common bean and pea, especially for developing durable resistance to stresses in crops. However, there is very limited information available about the release of commercial cultivars resulted from this strategy. Somers et al. implemented a molecular breeding strategy to introduce multiple pest resistance genes into Canadian wheat. They used high throughput SSR genotyping and half-seed analysis to process backcrossing and selection for six FHB resistance QTLs, plus orange blossom wheat midge resistance gene Sm1 and leaf rust resistance gene Lr21. They also used 45-76 SSR markers to perform background selection in backcrossing populations to accelerate the restoration of the RP genetic background. This strategy resulted in 87% fixation of the elite genetic background at the BC2F1 on average and successfully introduced all (up to 4) of the chromosome segments containing FHB, Sm1 and Lr21 resistance genes in four separate crosses. Joshi and Nayak and Xu recently reviewed the techniques and practical cases in marker-based gene pyramiding.

Similar to the simultaneous/synchronized backcrossing scheme, marker-assisted complex or convergent crossing (MACC) can be undertaken to pyramid multiple genes/QTLs. In particular, MACC is a proper option of breeding schemes for gene pyramiding if all the parents are improved cultivars or lines with good comprehensive performance and have different or complementary genes or favorable alleles for the traits of interest. The difference from simultaneous backcrossing is that selfing hybrid and progenies replaces backcrossing hybrid to the recurrent parent. In MACC, the hybrid of convergent crossing is subsequently self-pollinated and marker-based selection for target traits is performed for several consecutive generations until genetically stable lines with desired marker alleles and traits have been developed. In order to reduce population size and to avoid loss of most important genes/QTLs, different markers may be used and selected in different generations, depending on their relative importance. The markers for the most important genes/QTLs can be detected and selected first in early generations and less important markers later. Once homozygous alleles of the markers for a gene/locus are detected, they may not be necessarily detected again in the subsequent generations. Instead, phenotypic evaluation should be conducted if conditions permit.

Using markers to select or pyramid for multiple genes/QTLs is more complex and less proven. Recurrent selection is widely regarded as an effective strategy for the improvement of polygenic traits. However, the effectiveness and efficiency of selection are not so satisfactory in some cases because phenotypic selection is highly dependent upon environments and genotypic selection takes a longer time (2-3 crop seasons at least for one cycle of selection). Marker-assisted recurrent selection (MARS) is a scheme which allows performing genotypic selection and intercrossing in the same crop season for one cycle of selection. Therefore, MARS could enhance the efficiency of recurrent selection and accelerate the progress of the procedure, particularly helps in integrating multiple favorable genes/QTLs from different sources through recurrent selection based on a multiple-parental population.

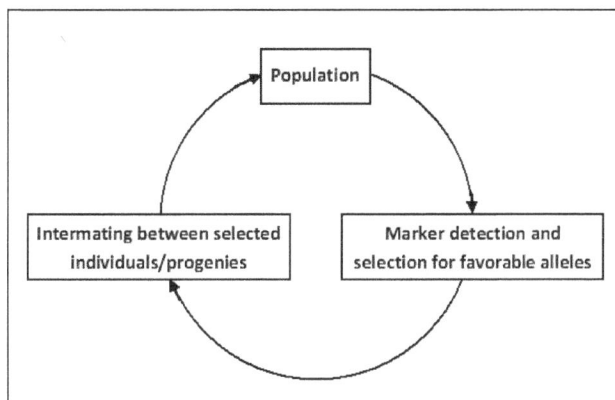

General procedure of marker-assisted recurrent selection (MARS).

For complex traits such as grain yield, biotic and abiotic resistance, MARS has been proposed for "forward breeding" of native genes and pyramiding multiple QTLs. As defined by Ribaut et al., MARS is a recurrent selection scheme using molecular markers for the identification and selection of multiple genomic regions involved in the expression of complex traits to assemble the best-performing genotype within a single or across related populations. Johnson presented an example to demonstrate the efficiency of MARS for quantitative traits. In their maize MARS programs, a large-scale use of markers in bi-parental populations, first for QTL detection and then for MARS on yield (i.e. rapid cycles of recombination and selection based on associated markers for yield), could allow increased efficiency of long-term selection by increasing the frequency of favorable alleles. Eathington and Crosbie et al. also indicated that the genetic gain achieved through MARS in maize was about twice that of phenotypic selection (PS) in some reference populations. In upland cotton, Yi et al. reported significant effectiveness of MARS for resistance to Helicoverpa armigera. The mean levels of resistance in improved populations after recurrent selection were significantly higher than those of preceding populations.

Genomic Selection

Genomic selection (GS) or genome-wide selection (GWS) is a form of marker-based selection, which was defined by Meuwissen as the simultaneous selection for many (tens or hundreds of thousands of) markers, which cover the entire genome in a dense manner so that all genes are expected to be in linkage disequilibrium with at least some of the markers. In GS genotypic data (genetic markers) across the whole genome are used to predict complex traits with accuracy sufficient to allow selection on that prediction alone. Selection of desirable individuals is based on genomic estimated breeding value (GEBV), which is a predicted breeding value calculated using an innovative method based on genome-wide dense DNA markers. GS does not need significant testing and identifying a subset of markers associated with the trait. In other words, QTL mapping with populations derived from specific crosses can be avoided in GS. However, it does first need to develop GS models, i.e. the formulae for GEBV prediction. In this process (training phase), phenotypes and genome-wide genotypes are investigated in the training population (a subset of a population) to predict significant relationships between phenotypes and genotypes using statistical approaches. Subsequently, GEBVs are used for the selection of desirable individuals in the breeding phase, instead of the genotypes of markers used in traditional MAS. For accuracy of GEBV and GS, genome-wide genotype data is necessary and require high marker density in which all quantitative trait loci (QTLs) are in linkage disequilibrium with at least one marker.

GS can be possible only when high-throughput marker technologies, high-performance computing and appropriate new statistical methods become available. This approach has become feasible due to the discovery and development of large number of single nucleotide polymorphisms (SNPs) by genome sequencing and new methods to efficiently genotype large number of SNP markers. As suggested by Goddard and Hayes, the ideal method to estimate the breeding value from genomic data is to calculate the conditional mean of the breeding value given the genotype at each QTL. This conditional mean can only be calculated by using a prior distribution of QTL effects, and thus this should be part of the research to implement GS. In practice, this method of estimating breeding values is approximated by using the marker genotypes instead of the QTL genotypes, but the ideal method is likely to be approached more closely as more sequence and SNP data are obtained.

Since the application of GS was proposed by Meuwissen et al. to breeding populations, theoretical, simulation and empirical studies have been conducted, mostly in animals. Relatively speaking, GS in plants was less studied and large-scale empirical studies are not available in public sectors for plant breeding, but it has attracted more and more attention in recent years. Studies indicated that in all cases, accuracies provided by GS were greater than might be achieved on the basis of pedigree information alone. In oil palm, for a realistic yet relatively small population, GS was superior to MARS and PS in terms of gain per unit cost and time. The studies have demonstrated the advantages of GS, suggesting that GS would be a potential method for plant breeding and it could be performed with realistic sizes of populations and markers when the populations used are carefully chosen.

GS has been highlighted as a new approach for MAS in recent years and is regarded as a powerful, attractive and valuable tool for plant breeding. However, GS has not become a popular methodology in plant breeding, and there might be a far way to go before the extensive use of GS in plant breeding programs. The major reason might be the unavailability of sufficient knowledge of GS for practical use. Statistics and simulation discussed in terms of formulae in GS studies are most likely too specific and hard for plant breeders to understand and to use in practical breeding programs. From a plant breeder's point of view, GS can be practicable for a few breeding populations with a specific purpose, but may be impractical for a whole breeding program dealing with hundreds and thousands of crosses/populations at the same time. Therefore, GS must shift from theory to practice, and its accuracy and cost effectiveness must be evaluated in practical breeding programs to provide convincing empirical evidence and warrant a practicable addition of GS to a plant breeder's toolbox. Development of easily understandable formulae for GEBVs and user-friendly software packages for GS analysis is helpful in facilitating and enhancing the application of GS in plant breeding. Kumpatla et al. recently presented an overall review on the GS for plant breeding.

Marker-based Breeding and Conventional Breeding

Marker-assisted breeding became a new member in the family of plant breeding as various types of molecular markers in crop plants were developed during the 1980s and 1990s. The extensive use of molecular markers in various fields of plant science, e.g. germplasm evaluation, genetic mapping, map-based gene discovery, characterization of traits and crop improvement, has proven that molecular technology is a powerful and reliable tool in genetic manipulation of agronomically important traits in crop plants. Compared with conventional breeding methods, MAB has significant advantages.

MAB can allow selection for all kinds of traits to be carried out at seedling stage and thus reduce the time required before the phenotype of an individual plant is known. For the traits that are expressed at later developmental stages, undesirable genotypes can be quickly eliminated by MAS. This feature is particularly important and useful for some breeding schemes such as backcrossing and recurrent selection, in which crossing with or between selected individuals is required.

MAB can be not affected by environment, thus allowing the selection to be performed under any environmental conditions (e.g. greenhouse and off-season nurseries). This is very helpful for improvement of some traits (e.g. disease/pest resistance and stress tolerance) that are expressed only when favorable environmental conditions present. For low-heritability traits that are easily affected by environments, MAS based on reliable markers tightly linked to the QTLs for traits of interest can be more effective and produce greater progress than phenotypic selection.

MAB using co-dominance markers (e.g. SSR and SNP) can allow effective selection of recessive alleles of desired traits in the heterozygous status. No selfing or test crossing is needed to detect the traits controlled by recessive alleles, thus saving time and accelerating breeding progress.

For the traits controlled by multiple genes/QTLs, individual genes/QTLs can be identified and selected in MAB at the same time and in the same individuals, and thus MAB is particularly suitable for gene pyramiding. In traditional phenotypic selection, however, to distinguish individual genes/loci is problematic as one gene may mask the effect of additional genes.

Genotypic assays based on molecular markers may be faster, cheaper and more accurate than conventional phenotypic assays, depending on the traits and conditions, and thus MAB may result in higher effectiveness and higher efficiency in terms of time, resources and efforts saved.

The research and use of MAB in plants has continued to increase in the public and private sectors, particularly since 2000s. However, MAS and MABC were and are primarily constrained to simply-inherited traits, such as monogenic or oligogenic resistance to diseases/pests, although quantitative traits were also involved. The application of molecular markers in plant breeding has not achieved the results as expected previously in terms of extent and success (e.g. release of commercial cultivars). Collard and Mackill listed ten reasons for the low impact of MAS and MAB in general. Improvement of most agronomic traits that are of complicated inheritance and economic importance like yield and quality is still a great challenge for MAB including the newly developed GS. From the viewpoint of a plant breeder, MAB is not universally or necessarily advantageous. The application of molecular technologies to plant breeding is still facing the following drawbacks and/or challenges:

Not all markers are breeder-friendly. This problem may be solved by converting of non-breeder-friendly markers to other types of breeder-friendly markers (e.g. RFLP to STS, sequence tagged site, and RAPD to SCAR, sequence characterized amplified region).

Not all markers can be applicable across populations due to lack of marker polymorphism or reliable marker-trait association. Multiple mapping populations are helpful in understanding marker allelic diversity and genetic background effects. In addition, QTL positions and effects also need to be validated and re-estimated by breeders in their specific germplasm.

False selection may occur due to recombination between the markers and the genes/QTLs of interest. Use of flanking markers or more markers for the target gene/QTL can help.

Imprecise estimates of QTL locations and effects result in slower progress than expected. The efficiency of QTL detection is attributed to multiple factors, such as algorithms, mapping methods, number of polymorphic markers, and population type and size. High marker density fine mapping with large populations and well-designed phenotyping across multiple environments may provide more accurate estimates of QTL location and effects.

A large number of breeding programs have not been equipped with adequate facilities and conditions for a large-scale adoption of MAB in practice.

The methods and schemes of MAB must be easily understandable, acceptable and implementable for plant breeders, unless they are not designed for a large scale use in practical breeding programs.

Higher Startup Expenses and Labor Costs

With a long history of development, especially since the fundamental principles of inheritance were established in the late 19th and early 20th centuries, plant breeding has become an important component of agricultural science, which has features of both science and arts. Conventional breeding methodologies have extensively proven successful in development of cultivars and germplasm. However, subjective evaluation and empirical selection still play a considerable role in conventional breeding. Scientific breeding needs less experience and more science. MAB has brought great challenges, opportunities and prospects for conventional breeding. As a new member of the whole family of plant breeding, however, MAB, as transgenic breeding or genetic manipulation does, cannot replace conventional breeding but is and only is a supplementary addition to conventional breeding. High costs and technical or equipment demands of MAB will continue to be a major obstacle for its large-scale use in the near future, especially in the developing countries. Therefore, integration of MAB into conventional breeding programs will be an optimistic strategy for crop improvement in the future. It can be expected that the drawbacks of MAB will be gradually overcome, as its theory, technology and application are further developed and improved. This should lead to a wide adoption and use of MAB in practical breeding programs for more crop species and in more countries as well.

References

- Mudge, K.; Janick, J.; Scofield, S.; Goldschmidt, E. (2009). A History of Grafting (PDF). Horticultural Reviews. 35. Pp. 449–475. Doi:10.1002/9780470593776.ch9. ISBN 9780470593776

- Plant-breeding, strategies-for-enhancement-in-food-production, biology, guides: toppr.com, Retrieved 4 January, 2019

- Siebert, S; Döll, P; Hoogeveen, J; Faures, J M; Frenken, K; Feick, S (2005). "Development and validation of the global map of irrigation areas". Hydrol. Earth Syst. Sci. 9 (5): 535–47. Doi:10.5194/hess-9-535-2005

- Noel Kingsbury (2009). Hybrid: The History and Science of Plant Breeding. University of Chicago Press. P. 140. ISBN 9780226437057

- Hybridization, conventional-plant-breeding, genetic-intensification, sid: ag4impact.org, Retrieved 25 August, 2019

- B. Barnabás; B. Obert; G. Kovács (1999). "Colchicine, an efficient genome-doubling agent for maize (Zea mays L.) Microspores cultured in anthero". Plant Cell Reports. 18 (10): 858–862. Doi:10.1007/s002990050674

- Molecular-markers-and-marker-assisted-breeding-in-plants, plant-breeding-from-laboratories-to-fields, books: intechopen.com, Retrieved 31 July, 2019

4

Agricultural Soil Science

The branch of soil science that focuses on the study of edaphic conditions with respect to the production of food and fiber is known as agricultural soil science. Some of the techniques studied within this field are contour plowing and mulching. The diverse applications of agricultural soil science and these techniques have been thoroughly discussed in this chapter.

SOIL AS FOUNDATION OF AGRICULTURE

Soil is frequently referred to as the "fertile substrate", but not all soils are suitable for growing crops. Ideal soils for agriculture are balanced in contributions from mineral components (sand: 0.05–2 mm, silt: 0.002–0.05 mm, clay: <0.002 mm), soil organic matter (SOM), air, and water. The balanced contributions of these components allow for water retention and drainage, oxygen in the root zone, nutrients to facilitate crop growth; and they provide physical support for plants. The distribution of these soil components in a particular soil is influenced by the five factors of soil formation: parent material, time, climate, organisms, and topography. Each one of these factors plays a direct and overlapping role in influencing the suitability of a soil for agriculture.

Inorganic Soil Components

As one might expect, contributions from each mineral size fraction help to provide the physical framework for a productive soil. Loamy-textured soils are commonly described as medium textured with functionally-equal contributions of sand, silt, and clay. These medium-textured soils are often considered ideal for agriculture as they are easily cultivated by farmers and can be highly productive for crop growth.

The mineral components of soil may exist as discrete particles, but are more commonly associated with one another in larger aggregates that provide structure to soil. These aggregates, or peds, play an important role in influencing the movement of water and air through soil. Sandy soils have large pore spaces and increase water drainage, but do not provide soils with many nutrients. Clay-rich soils, on the other hand, increase water holding capacity and provide many plant essential nutrients. A common measure of soil fertility is obtained by measuring the cation exchange capacity (CEC). The CEC is a measure of a soil's ability to exchange positive ions between the soil particles and solution surrounding these particles.

Due to their high surface area, clay particles can exert a large influence on various soil properties (e.g., CEC, structure, water-holding capacity), even when the percent clay content is low. Clay minerals are colloidal particles, having high surface area, with charged surfaces; permitting binding of many essential plant nutrients. The most prevalent clay-sized particles in soils fall into the class of layer-type aluminosilicates that commonly have permanent negative charge with a high CEC. Positively charged clay particles, which bind anions, include those which have pH dependent charge. The most common classes of these minerals in soils are the iron (Fe), aluminum (Al), and manganese (Mn) (hydr)oxides.

Soil Organic Matter (SOM)

SOM comprises the partial or well-decomposed residues of organic biomass present in soil. SOM gives topsoil its deep black colors and rich aromas that many home gardeners and farmers of grassland soils are familiar with. Surface soils are composed of approximately 1 to 6% organic matter, with SOM decreasing with depth. The 'Great Plains' of North America and the 'Bread Basket' of Europe are some of the world's most productive agricultural soils because they developed under grassland vegetation, whose root biomass and decomposition resulted in SOM accumulation. Figure is a photograph of an organic matter-rich soil (mollisol) formed under prairie vegetation in the United States. The thick dark upper layers in this soil reflect the high SOM content. The presence of SOM is crucial for fertile soil as it provides essential plant nutrients, beneficially influences soil structure, buffers soil pH, and improves water holding capacity and aeration. The presence of organic, ionizable functional groups (e.g., carboxyl, alcoholic/phenolic OH, enol, quinone, and amine) impart charge to SOM, contributing high CEC, and pH buffering capacity.

Figure: Mollisol soil profile showing thick dark horizon
with high organic matter content.

Soil pH

Often referred to as the master variable of soil, pH controls a wide range of physical, chemical, and biological processes and properties that affect soil fertility and plant growth. Soil pH, which reflects the acidity level in soil, significantly influences the availability of plant nutrients, microbial activity, and even the stability of soil aggregates. At low pH, essential plant macronutrients (i.e., N, P, K, Ca, Mg, and S) are less bioavailable than at higher pH values near 7, and certain micronutrients (i.e., Fe, Mn, Zn) tend to become more soluble and potentially toxic to plants at low pH values (5–6). Aluminum toxicity is also a common problem for crop growth at low pH (<5.5). Typically, soil pH values from 6 to 7.5 are optimal for plant growth; however, there are certain plants species that can tolerate — or even prefer — more acidic or basic conditions. Maintaining a narrow range in soil pH is beneficial to crop growth. SOM and clay minerals help to buffer soils to maintain a pH range optimal for plant growth. In instances where the pH is outside a desirable range, the soil pH can be altered through amendments such as lime to raise the pH. Ammonium sulfate, iron sulfate, or elemental sulfur can be added to soil to lower pH.

Soil Degradation and Crop Production

Soil forms from fresh parent material through various chemical and physical weathering processes and SOM is incorporated into soil through decomposition of plant residues and other biomass. Although these natural soil building processes regenerate the soil, the rate of soil formation is very slow. For this reason, soil should be considered a nonrenewable resource to be conserved with care for generations to come. The rate of soil formation is hard to determine and highly variable, based on the five factors of soil formation. Scientists have calculated that 0.025 to 0.125 mm of soil is produced each year from natural soil forming processes. Because of the time required to generate new soil, it is imperative that agricultural practices utilize best management practices (BMPs) to prevent soil erosion. The soil which is first eroded is typically the organic and nutrient enriched surface layer which is highly beneficial for plant growth. Thus, the primary on-site outcome is reduced crop yield as only the less fertile subsurface layers remain. Soil erosion also pollutes adjacent streams and waterways with sediment, nutrients, and agrochemicals creating serious off-site impacts.

Historically, conventional agriculture has accelerated soil erosion to rates that exceed that of soil formation (Table). Erosion is often accelerated by agricultural practices that leave the soil without adequate plant cover and therefore exposed to raindrop splash and surface runoff or wind. Throughout human history, soil erosion has affected the ability of societies to produce an adequate food supply. Poignant examples of this can be seen in the eroded silt built up in the ancient riverbeds of Mesopotamia, making irrigation problematic, and the United States Dust Bowl of the 1930s where a devastating drought increased wind erosion, carrying fertile topsoil from the Midwest hundreds of kilometers to Washington, DC. Figure is a stunning photograph demonstrating the devastating effects of this severe wind erosion. The Dust Bowl made soil erosion a high priority in the American public consciousness of the 1930s, and it remains a top priority today.

Measurement type	Sample size n	Mean rate of soil erosion (mm/yr)	Net rate of soil formation (mm/yr)
Conventional Agriculture	448	1.54 (0.32)[‡]	-1.52 to -1.42
Conservation Agriculture	47	0.082 (0.022)	-0.057 to +0.043
Native Vegetation	65	0.013 (0.016)	+0.012 to +0.112
Geological	925	0.029 (0.029)	-4.00×10^{-3} to +0.096

Figure: Tractor and farm equipment buried by soil transported by wind.

Today, agricultural fields are not immune to the forces of nature (e.g., moving water, blowing wind, extremes of temperature) that caused soil erosion in the past. Figure shows the severe effects of surface runoff and soil loss in the northwestern United States. Implementation of agricultural best management practices (BMPs), and through the practice of conservation agriculture, the rate of soil loss can be reduced to approximately equal the rate of soil formation, although often still greater than that in natural systems (table). In addition to soil erosion, intensive land use has resulted in deforestation, water shortages, and rapidly increasing desertification of vast areas of the globe, all of which threaten the sustainability of our agricultural systems.

Figure: Damage to agricultural field in Washington
(United States) resulting from water erosion.

Sustainable Soil Management

It is evident that, in order to maintain and increase food production, efforts to prevent soil degradation must become a top priority of our global society. Current population models predict a global population of between 8 and 10 billion in the next 50 years and a two-fold increase in food demand. If mismanagement of soil resources continues to diminish the fertility of the soil and the amount of productive arable land, then we will have lost a precious and essential pillar of sustainable agriculture. Sustainable agriculture is an approach to farming that focuses on production of food in a manner that can be maintained with minimal degradation of ecosystems and natural resources. This sustainable approach to agriculture strives to protect environmental resources, including soil, and provide economic profitability while maintaining social equity. The concept of sustainable agriculture is often misinterpreted to mean that chemical fertilizers and pesticides should never be used. This notion is incorrect, as sustainable agriculture should embrace those practices that provide the most beneficial services for agroecosystems and encourage long-term production of food supplies in a cultural context of the region. It cannot be overstressed that sustainable practices should not only consider crop production and profit, but must include land management strategies that reduce soil erosion and protect water resources. By embracing certain modern-day technologies, proven BMPs, and learning from the past, our society will be able to continue to conserve soil resources and produce food supplies sufficient to meet current and future population demands.

Soil Fertility and Crop Growth

The early use of fire to flush out wild game and to clear forested land provided the first major anthropogenic influence on the environment. By burning native vegetation, early humans were able to gain access to herbivores grazing on the savanna and in nearby woodlands, and to suppress the growth of less desirable plant species for those easier to forage and eat. These and other factors (e.g., population pressures, climate change, encouraging/protecting desirable plants), help to lay the groundwork for the Agricultural Revolution and caused a dramatic shift in the interactions between humans and the earth. The shift from hunter-gatherer societies to an agrarian way of life drastically changed the course of human history and irreversibly altered natural nutrient cycling within soils. When humans sowed the first crop seeds at the dawn of the Neolithic Period, the soil provided plant-essential nutrients and served as the foundation for human agriculture.

Plant Nutrients

Throughout Earth's history, natural cycling of nutrients has occurred from the soil to plants and animals, and then back to the soil, primarily through decomposition of biomass. This cycling helps to maintain the essential nutrients required for plant growth in the soil. Complex nutrient cycles incorporate a range of physical, chemical, and — most importantly — biological processes to trace the fate of specific plant nutrients (e.g., N, P, C, S) in the environment. For a thorough analysis of these cycles, additional reference materials are available. a simplified version of nutrient cycling in natural and agricultural systems is shown in figure.

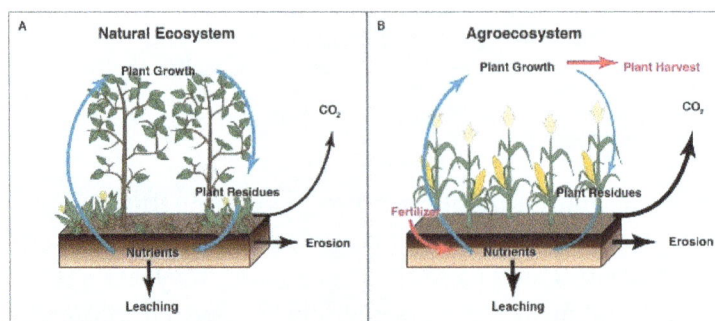

Simplified nutrient cycling schemes for (a) a natural ecosystem and (b) an agroecosystem.

The thickness of arrows correspond to the relative amounts. The blue arrows represent basic nutrient pathways such as plant uptake, biomass decomposition, and nutrient return to soil. In the agroecosystem fertilizer (e.g., manure, compost, chemical) is applied and nutrients are removed through plant harvest (red arrows). Greater potential leaching (e.g., nitrate), erosion (e.g., soil, phosphate), and CO2 emissions in the agroecosystem are indicated via thick black arrows.

It is generally accepted that there are 17 essential elements required for plant growth. The lack of any one of these essential nutrients, listed in Table, can result in a severe limitation of crop yield — an example of the principle of limiting factors. Of the mineral elements, the primary macronutrients (N, P, and K) are needed in the greatest quantities from the soil and are the plant nutrients most likely to be in short supply in agricultural soils. Secondary macronutrients are needed in smaller quantities, are typically in sufficient quantities in soil, and therefore are not often limiting for crop growth. The micronutrients, or sometimes called trace nutrients, are needed in very small amounts and, if in excess, can be toxic to plants. Silicon (Si) and sodium (Na) are sometimes considered to be essential plant nutrients, but due to their ubiquitous presence in soils they are never in short supply.

Table: Essential plant nutrient elements and their primary form utilized by plants.

Essential plant element		Symbol	Primary form
Non-Mineral Elements			
	Carbon	C	CO_2 (g)
	Hydrogen	H	H_2O (l), H^+
	Oxygen	O	H_2O (l), O_2 (g)
Mineral Elements			
Primary Macronutrients	Nitrogen	N	NH_4^+, NO_3^-
	Phosphorus	P	HPO_4^{2-}, $H_2PO_4^-$
	Potassium	K	K^+
Secondary Macronutrients	Calcium	Ca	Ca^{2+}
	Magnesium	Mg	Mg^{2+}
	Sulfur	S	SO_4^{2-}
Micronutrients	Iron	Fe	Fe^{3+}, Fe^{2+}
	Manganese	Mn	Mn^{2+}
	Zinc	Zn	Zn^{2+}
	Copper	Cu	Cu^{2+}
	Boron	B	$B(OH)_3$
	Molybdenum	Mo	MoO_4^{2-}
	Chlorine	Cl	Cl^-
	Nickel	Ni	Ni^{2+}

Agriculture alters the natural cycling of nutrients in soil. Intensive cultivation and harvesting of crops for human or animal consumption can effectively mine the soil of plant nutrients. In order to maintain soil fertility for sufficient crop yields, soil amendments are typically required. Early humans soon learned to amend their fields with animal manure, charcoal, ash, and lime ($CaCO_3$) to improve soil fertility. Today, farmers add numerous soil amendments to enhance soil fertility, including inorganic chemical fertilizers and organic sources of nutrients, such as manure or compost, often resulting in surplus quantities of primary macronutrients. The efficiency of fertilizer application and use by crops is not always optimized, and excess nutrients, especially N and P, can be transported via surface runoff or leaching from agricultural fields and pollute surface- and groundwater.

AGRICULTURAL SOIL SCIENCE

Agricultural soil science is a branch of soil science that deals with the study of edaphic conditions as they relate to the production of food and fiber. In this context, it is also a constituent of the field of agronomy and is thus also described as soil agronomy.

Agricultural soil science follows the holistic method. Soil is investigated in relation to and as integral part of terrestrial ecosystems but is also recognized as a manageable natural resource.

Agricultural soil science studies the chemical, physical, biological, and mineralogical composition of soils as they relate to agriculture. Agricultural soil scientists develop methods that will improve the use of soil and increase the production of food and fiber crops. Emphasis continues to grow on the importance of soil sustainability. Soil degradation such as erosion, compaction, lowered fertility, and contamination continue to be serious concerns. They conduct research in irrigation and drainage, tillage, soil classification, plant nutrition, soil fertility, and other areas.

Although maximizing plant (and thus animal) production is a valid goal, sometimes it may come at high cost which can be readily evident (e.g. massive crop disease stemming from monoculture) or long-term (e.g. impact of chemical fertilizers and pesticides on human health). An agricultural soil scientist may come up with a plan that can maximize production using sustainable methods and solutions, and in order to do that he must look into a number of science fields including agricultural science, physics, chemistry, biology, meteorology and geology.

Kinds of Soil and their Variables

Some soil variables of special interest to agricultural soil science are:

- Soil texture or soil composition: Soils are composed of solid particles of various sizes. In decreasing order, these particles are sand, silt and clay. Every soil can be classified according to the relative percentage of sand, silt and clay it contains.

- Aeration and porosity: Atmospheric air contains elements such as oxygen, nitrogen, carbon and others. These elements are prerequisites for life on Earth. Particularly, all cells (including root cells) require oxygen to function and if conditions become anaerobic they fail to respire and metabolize. Aeration in this context refers to the mechanisms by which air is delivered to the soil. In natural ecosystems soil aeration is chiefly accomplished through the vibrant activity of the biota. Humans commonly aerate the soil by tilling and plowing, yet such practice may cause degradation. Porosity refers to the air-holding capacity of the soil.

- Drainage: In soils of bad drainage the water delivered through rain or irrigation may pool and stagnate. As a result, prevail anaerobic conditions and plant roots suffocate. Stagnant water also favors plant-attacking water molds. In soils of excess drainage, on the other hand, plants don't get to absorb adequate water and nutrients are washed from the porous medium to end up in groundwater reserves.

- Water content: Without soil moisture there is no transpiration, no growth and plants wilt. Technically, plant cells lose their pressure. Plants contribute directly to soil moisture. For instance, they create a leafy cover that minimizes the evaporative effects of solar radiation. But even when plants or parts of plants die, the decaying plant matter produces a thick organic cover that protects the soil from evaporation, erosion and compaction.

- Water potential: Water potential describes the tendency of the water to flow from one area of the soil to another. While water delivered to the soil surface normally flows downward due to gravity, at some point it meets increased pressure which causes a reverse upward flow. This effect is known as water suction.

- Horizonation: Typically found in advanced and mature soils, horizonation refers to the creation of soil layers with differing characteristics. It affects almost all soil variables.

- Fertility: A fertile soil is one rich in nutrients and organic matter. Modern agricultural methods have rendered much of the arable land infertile. In such cases, soil can no longer support on its own plants with high nutritional demand and thus needs an external source of nutrients. However, there are cases where human activity is thought to be responsible for transforming rather normal soils into super-fertile ones.

- Biota and soil biota: Organisms interact with the soil and contribute to its quality in innumerable ways. Sometimes the nature of interaction may be unclear, yet a rule is becoming evident: The amount and diversity of the biota is "proportional" to the quality of the soil. Clades of interest include bacteria, fungi, nematodes, annelids and arthropods.

- Soil acidity or soil pH and cation-exchange capacity: Root cells act as hydrogen pumps and the surrounding concentration of hydrogen ions affects their ability to absorb nutrients. pH is a measure of this concentration. Each plant species achieves maximum growth in a particular pH range, yet the vast majority of edible plants can grow in soil pH between 5.0 and 7.5.

Soil scientists use a soil classification system to describe soil qualities. The International Union of Soil Sciences endorses the World Reference Base as the international standard.

Soil Preservation

In addition, agricultural soil scientists develop methods to preserve the agricultural productivity of soil and to decrease the effects on productivity of erosion by wind and water. For example, a technique called contour plowing may be used to prevent soil erosion and conserve rainfall. Researchers in agricultural soil science also seek ways to use the soil more effectively in addressing associated challenges. Such challenges include the beneficial reuse of human and animal wastes using agricultural crops; agricultural soil management aspects of preventing water pollution and the build-up in agricultural soil of chemical pesticides. Regenerative agriculture practices can be used to address these challenges and rebuild soil health.

INTEGRATED NUTRIENT MANAGEMENT

Integrated nutrient management is the combined application of chemical fertilizers along with organic resource materials like, organic manures, green manures, bi-fertilizers and other organic decomposable materials for crop production.

The basic concepts of IPNS is the maintenance or adjustment of soil fertility and supply of plant nutrients to an optimum level for sustaining desired crop productivity through optimization of benefits from all possible sources of plant nutrients in an integrated manner.

IPNS is ecologically, socially and economically viable and environment friendly which can be practiced by farmers to derive higher productivity with simultaneously maintaining soil fertility. Integrated nutrient management encourages the use of on-farm organics, thus it saves on the cost of fertilizers for crop production.

The basic concept of integrated nutrient management (INM) or integrated plant nutrition management (IPNM) is the adjustment of plant nutrient supply to an optimum level for sustaining the desired crop productivity.

It involves proper combination of chemical fertilizers, organic manure, crop, residues, N2~fixing crops (like pulses such as rice bean, Black gram, other pulses and oilseeds such as soybean and bio-fertilizers suitable to the system of land use and ecological, social and economic conditions.

The cropping system rather than an individual crop, and farming system rather than an individual field, is the focus of attention in this approach for development INM practices for various categories. The basic concept of INM is the maintenance of soil fertility, sustainable agricultural productivity and improving profitability through judicious and efficient use of fertilizers as mentioned.

Concepts of INM

INM use five major sub-concepts, viz:

1. Plant nutrients stored in the soil.

2. Plant nutrients, those present in the crop residues, organic manure and domestic wastes.

3. Plant nutrients purchased or obtained from outside the farm.

4. Plant nutrient looses e.g. those removed from the field in crop harvest and lost from the soil through volatilization (ammonia and nitrogen oxide gases and leaching (nitrate, sulphate etc).

5. Plant nutrient outputs e.g. nutrient uptake by the crops at harvest time.

The integrated plant nutrition management on integrated nutrient management thus firstly operates at plot level, optimizing the utilization of plant nutrients from diverse sources which are locally available, in order to improve the agronomic efficiency of such nutrient and at the same time reducing the losses of nutrients. Organic recycling is especially promoted through INM for facilitating nutrient recycling to improve soil fertility and productivity as low cost.

Goals of INM

To Maintain Soil Poductivity: To ensure productive and sustainable agriculture. To reduce expenditure on costs of purchased inputs by using farm manure and crop residue, etc. To utilize the potential benefits of green manures, leguminous crops and bio-fertilizers. To prevent degradation of the environment. To meet the social and economic aspiration of the farmers without harming the natural resource base of agricultural production.

Implementation of INM Activities

Different stages of implementation of INM are as follows:

1. Diagnosis phase: collection of background information.

2. Analysis of constraints.

3. Preparing potentiality and feasibility summary.

4. On-farm demonstrations.

5. Evaluation of INM activities.

Components of INM

1. Integration of soil fertility restoring crops like green manures, legumes etc.

2. Recycling of crop residues.

3. Use of organic manures like FYM, compost, vermicompost, biogas, slurry, poultry manure, bio-compost, press mud cakes, phosphocompost.

4. Utilization of Bio fertilizers.

5. Efficient genotypes and lastly.

6. Balanced use of fertilizer nutrients as per the requirement and target yields.

Diagnostic Phase of INM

In the first stage of diagnostic phase information with regard to the following is collected and analysed:

1. Farming/cropping systems.

2. Crop varieties grown.

3. Awareness about soil fertility problems.

4. Use of chemical fertilizers, lime/dolomite and other agro-chemicals.

5. Use of organic manure.

6. Availability of fertilizer and other inputs.

7. Irrigation sources and practices.

8. Soil testing service facility.

9. Constraints in the adoption of INM technologies.

10. Consideration of market opportunities.

The INM technologies must be compatible with the local farming system if they are to find acceptance and adoption. Therefore, attention must be paid to examine the interaction among different components of INM and the management of crops and animals that form the farming system.

Some of the agronomic parameters which need attention are cropping pattern, intercropping practices, biological condition of the field (weeds, diseases, and insects), soil conditions, irrigation facilities and climatic conditions.

Common Constraints of INM

Common constraints encountered by the farmers in adoption of INM technology are as follows:

1. Non-availability of FYM.

2. Difficulties in growing green manure crops.

3. Non-availability of bio-fertilizers.

4. Non-availability of soil testing facilities.

5. High cost of chemical fertilizers.

6. Non-availability of water.

7. Lack of knowledge and poor advisory services.

8. Non-availability of improved seeds.

Soil Texture

Soil texture is a classification instrument used both in the field and laboratory to determine soil classes based on their physical texture. Soil texture can be determined using qualitative methods such as texture by feel, and quantitative methods such as the hydrometer method. Soil texture has agricultural applications such as determining crop suitability and to predict the response of the soil to environmental and management conditions such as drought or calcium (lime) requirements. Soil texture focuses on the particles that are less than two millimeters in diameter which include sand, silt, and clay. The USDA soil taxonomy and WRB soil classification systems use 12 textural classes whereas the UK-ADAS system uses 11. These classifications are based on the percentages of sand, silt, and clay in the soil.

Soil Texture Classification

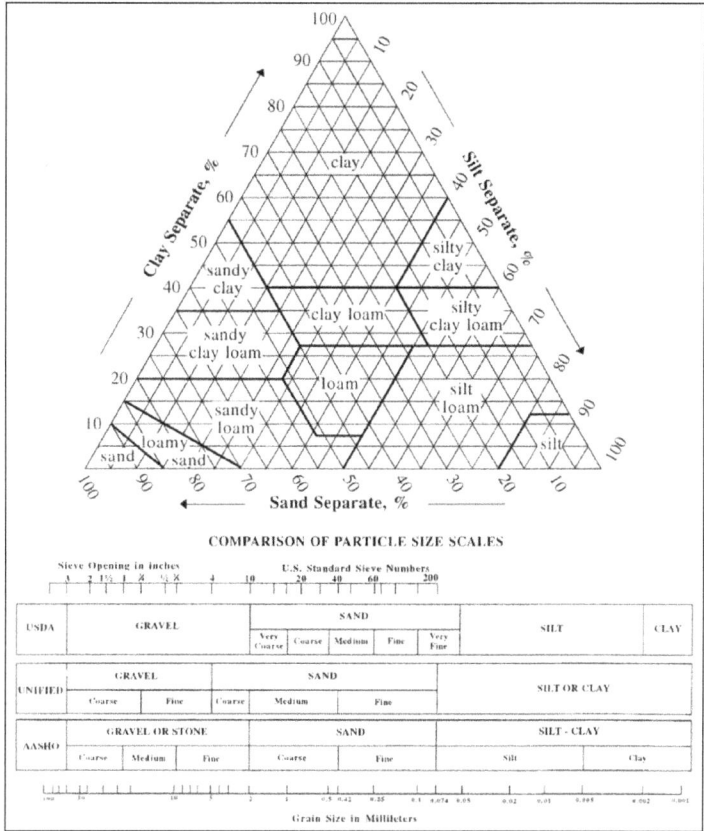

Soil texture triangle, showing the 12 major textural
classes, and particle size scales.

In the United States, twelve major soil texture classifications are defined by the USDA. The twelve classifications are sand, loamy sand, sandy loam, loam, silt loam, silt, sandy clay loam, clay loam, silty clay loam, sandy clay, silty clay, and clay. Soil textures are classified by the fractions of each soil separate (sand, silt, and clay) present in a soil. Classifications are typically named for the primary constituent particle size or a combination of the most abundant particles sizes, e.g. "sandy clay" or "silty clay". A fourth term, loam, is used to describe equal properties of sand, silt, and clay in a soil sample, and lends to the naming of even more classifications, e.g. "clay loam" or "silt loam".

Determining soil texture is often aided with the use of a soil texture triangle. One side of the triangle represents percent sand, the second side represents percent clay, and the third side represents percent silt. If you know the percentages of sand, clay, and silt in your soil sample, the triangle can be used to determine which of the twelve soil types you have. To do this, find your percentage of sand along the bottom of the triangle. Then follow the slanted line up to the left until you reach your percentage of clay. Where that point is will tell you what soil type you have. For example, if your soil is 70 percent sand and 10 percent clay then your soil is classified as a sandy loam. The same method can be used starting on any side of the soil triangle. If the texture by feel method was used to determine which type of soil you had, the triangle can also provide a rough estimate on the percentages of sand, silt, and clay in your soil.

Chemical and physical properties of a soil are related to texture. Particle size and distribution will affect a soil's capacity for holding water and nutrients. Fine textured soils generally have a higher capacity for water retention, whereas sandy soils contain large pore spaces that allow leaching.

Soil Separates

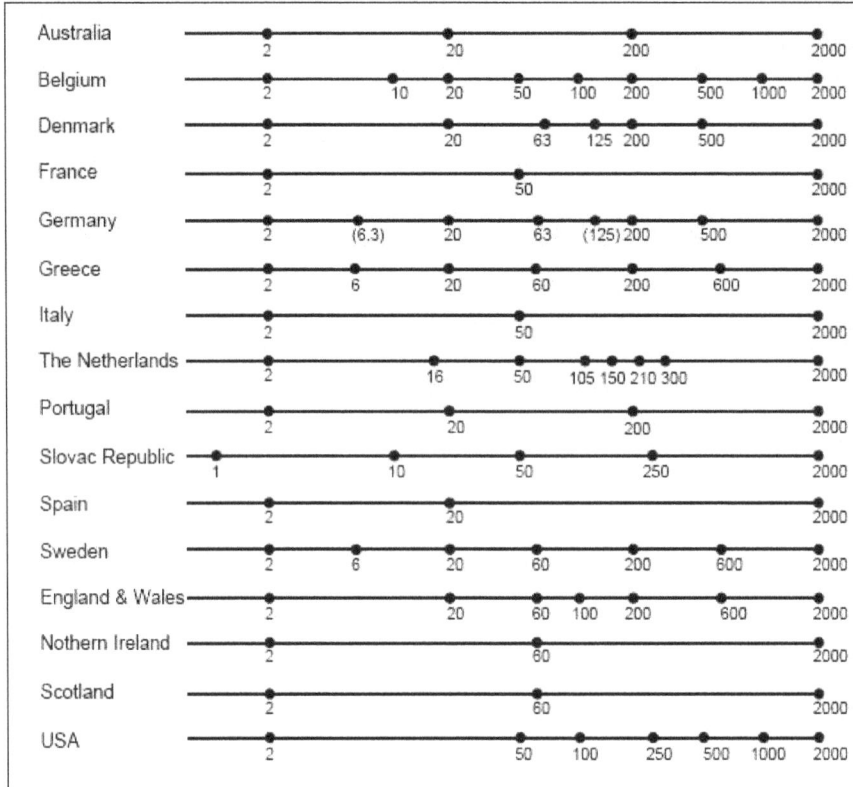

Particle size classifications used by different countries, diameters in μm

Soil separates are specific ranges of particle sizes. The smallest particles are *clay* particles and are classified as having diameters of less than 0.002 mm. Clay particles are plate-shaped instead of spherical, allowing for an increased specific surface area. The next smallest particles are *silt* particles and have diameters between 0.002 mm and 0.05 mm (in USDA soil taxonomy). The largest particles are *sand* particles and are larger than 0.05 mm in diameter. Furthermore, large sand particles can be described as *coarse*, intermediate as *medium*, and the smaller as *fine*. Other countries have their own particle size classifications.

Name of soil separate	Diameter limits (mm) (USDA classification)	Diameter limits (mm) (WRB classification)
Clay	less than 0.002	less than 0.002
Silt	0.002 – 0.05	0.002 – 0.063
Very fine sand	0.05 – 0.10	0.063 – 0.125
Fine sand	0.10 – 0.25	0.125 – 0.20
Medium sand	0.25 – 0.50	0.20 – 0.63
Coarse sand	0.50 – 1.00	0.63 – 1.25
Very coarse sand	1.00 – 2.00	1.25 – 2.00

Methods to Determine Soil Texture

Texture by Feel

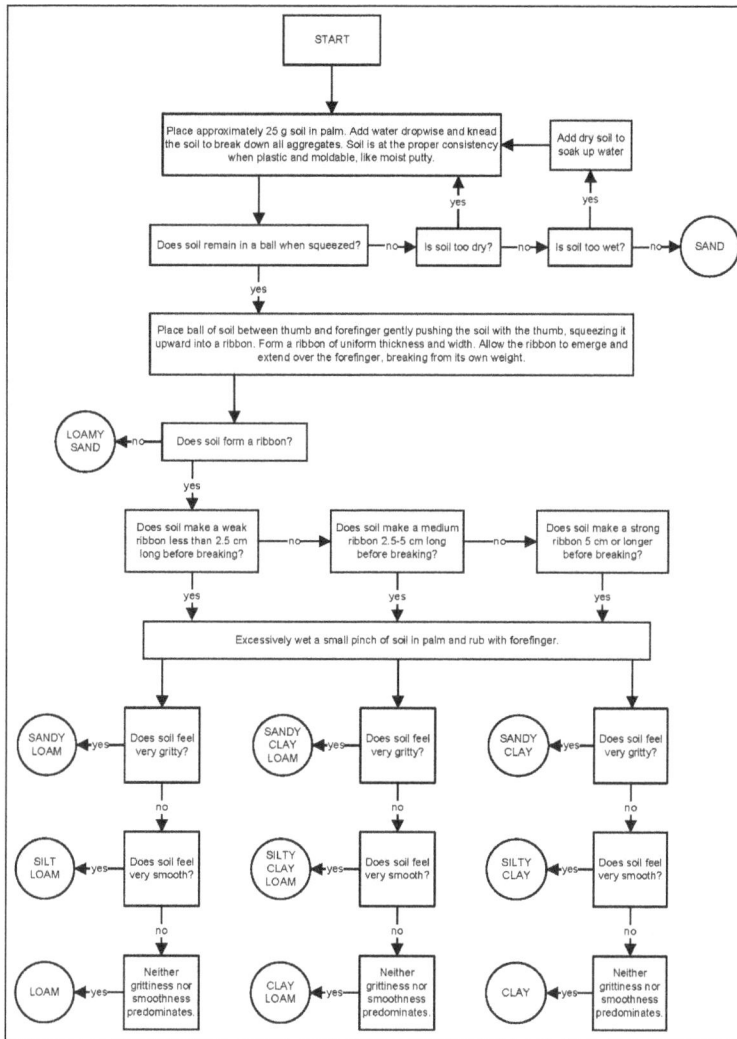

Texture by feel flow chart.

Hand analysis is a simple and effective means to rapidly assess and classify a soil's physical condition. Correctly executed, the procedure allows for rapid and frequent assessment of soil characteristics with little or no equipment. It is thus an extremely useful tool for identifying spatial variation both within and between fields as well as identifying progressive changes and boundaries between soil map units (soil series). Texture by feel is a qualitative method, it does not provide exact values of sand, silt, and clay. Although qualitative, the texture by feel flowchart can be an accurate way for a scientist or interested individual to analyze the relative proportions of sand, silt, and clay.

The texture by feel method involves taking a small sample of soil and making a ribbon. A ribbon can be made by taking a ball of soil and pushing the soil between your thumb and forefinger, squeezing it upward into a ribbon. Allow the ribbon to emerge and extend over the forefinger, breaking from its own weight. Measuring the length of the ribbon can help determine the amount of clay in the sample. After making a ribbon, excessively wet a small pinch of soil in the palm of your hand and rub in with your forefinger to determine the amount of sand in the sample. Soils that have a high percentage of sand, such as sandy loam, or sandy clay, have a gritty texture. Soils that have a high percentage of silt, such as silty loam or silty clay, feel smooth. Soils that have a high percentage of clay, such as clay loam, have a sticky feel. Although the texture by feel method takes practice, it is a useful way to determine soil texture, especially in the field.

Hydrometer Method

The hydrometer method of determining soil texture is a quantitative measurement providing estimates of the percent sand, clay, and silt in the soil. The hydrometer method was developed in 1927 and is still widely used today. This method requires a chemical compound, sodium hexametaphsophate, which acts as a dispersing agent to separate soil aggregates. The soil is mixed with the sodium hexametaphosphate solution on an orbital shaker overnight. The solution is transferred to one liter graduated cylinders and filled with water. The soil solution is mixed with a metal plunger to disperse the soil particles. The soil particles separate based on size and sink to the bottom. Sand particles are the largest (2.00 – 0.05 mm) and sink to the bottom of the cylinder first. Silt particles are the medium-sized (0.05 – 0.002 mm) and sink to the bottom of the cylinder after the sand. Clay particles are the smallest (<0.002 mm) and separate out above the silt layer. Measurements are taken using a soil hydrometer. A soil hydrometer measures that relative density of liquids (density of a liquid to the density of water). The hydrometer is lowered into the cylinder containing the soil mixture at different times, forty-five seconds to measure sand content, one and a half hours to measure silt content and between six and twenty-four hours (depending on the protocol used) to measure clay. The number on the hydrometer that is visible (above the soil solution) is recorded. A blank (containing only water and the dispersing agent) is used to calibrate the hydrometer.

The values recorded from the readings are used to calculate the percent clay, silt and sand. The blank is subtracted from each of the three readings. The calculations are as follows:

- Percent silt = (dried mass of soil – (sand hydrometer reading – blank reading)/(dried mass of soil) × 100.

- Percent clay = (clay hydrometer reading – blank reading)/(dried mass of soil) × 100.

- Percent sand = 100 – (percent clay + percent silt).

Additional Methods

There are several additional quantitative methods to determine soil texture. Some examples of these methods are the pipette method, the particulate organic matter (POM) method, and the rapid method.

Soil Fertility

Soil fertility refers to the ability of soil to sustain agricultural plant growth, i.e. to provide plant habitat and result in sustained and consistent yields of high quality. A fertile soil has the following properties:

- The ability to supply essential plant nutrients and water in adequate amounts and proportions for plant growth and reproduction; and

- The absence of toxic substances which may inhibit plant growth.

The following properties contribute to soil fertility in most situations:

- Sufficient soil depth for adequate root growth and water retention;

- Good internal drainage, allowing sufficient aeration for optimal root growth (although some plants, such as rice, tolerate waterlogging);

- Topsoil with sufficient soil organic matter for healthy soil structure and soil moisture retention;

- Soil pH in the range 5.5 to 7.0 (suitable for most plants but some prefer or tolerate more acid or alkaline conditions);

- Adequate concentrations of essential plant nutrients in plant-available forms;

- Presence of a range of microorganisms that support plant growth.

In lands used for agriculture and other human activities, maintenance of soil fertility typically requires the use of soil conservation practices. This is because soil erosion and other forms of soil degradation generally result in a decline in quality with respect to one or more of the aspects indicated above.

Soil Fertilization

Bioavailable phosphorus is the element in soil that is most often lacking. Nitrogen and potassium are also needed in substantial amounts. For this reason these three elements are always identified on a commercial fertilizer analysis. For example, a 10-10-15 fertilizer has 10 percent nitrogen, 10 percent (P_2O_5) available phosphorus and 15 percent (K_2O) water-soluble potassium. Sulfur is the fourth element that may be identified in a commercial analysis—e.g. 21-0-0-24 which would contain 21% nitrogen and 24% sulfate.

Inorganic fertilizers are generally less expensive and have higher concentrations of nutrients than organic fertilizers. Also, since nitrogen, phosphorus and potassium generally must be in the inorganic forms to be taken up by plants, inorganic fertilizers are generally immediately bioavailable to plants without modification. However, some have criticized the use of inorganic fertilizers, claiming that the water-soluble nitrogen doesn't provide for the long-term needs of the plant and creates water pollution. Slow-release fertilizers may reduce leaching loss of nutrients and may make the nutrients that they provide available over a longer period of time.

Soil fertility is a complex process that involves the constant cycling of nutrients between organic and inorganic forms. As plant material and animal wastes are decomposed by micro-organisms, they release inorganic nutrients to the soil solution, a process referred to as mineralization. Those nutrients may then undergo further transformations which may be aided or enabled by soil micro-organisms. Like plants, many micro-organisms require or preferentially use inorganic forms of nitrogen, phosphorus or potassium and will compete with plants for these nutrients, tying up the nutrients in microbial biomass, a process often called immobilization. The balance between immobilization and mineralization processes depends on the balance and availability of major nutrients and organic carbon to soil microorganisms. Natural processes such as lightning strikes may fix atmospheric nitrogen by converting it to (NO_2). Denitrification may occur under anaerobic conditions (flooding) in the presence of denitrifying bacteria. Nutrient cations, including potassium and many micronutrients, are held in relatively strong bonds with the negatively charged portions of the soil in a process known as cation exchange.

In 2008 the cost of phosphorus as fertilizer more than doubled, while the price of rock phosphate as base commodity rose eight-fold. Recently the term peak phosphorus has been coined, due to the limited occurrence of rock phosphate in the world.

Light and CO_2 Limitations

Photosynthesis is the process whereby plants use light energy to drive chemical reactions which convert CO_2 into sugars. As such, all plants require access to both light and carbon dioxide to produce energy, grow and reproduce.

While typically limited by nitrogen, phosphorus and potassium, low levels of carbon dioxide can also act as a limiting factor on plant growth. Peer-reviewed and published scientific studies have shown that increasing CO_2 is highly effective at promoting plant growth up to levels over 300 ppm. Further increases in CO_2 can, to a very small degree, continue to increase net photosynthetic output.

Soil Depletion

Soil depletion occurs when the components which contribute to fertility are removed and not replaced, and the conditions which support soil's fertility are not maintained. This leads to poor crop yields. In agriculture, depletion can be due to excessively intense cultivation and inadequate soil management.

Soil fertility can be severely challenged when land use changes rapidly. For example, in Colonial New England, colonists made a number of decisions that depleted the soils, including: allowing herd animals to wander freely, not replenishing soils with manure, and a sequence of events that led to erosion. William Cronon wrote that "the long-term effect was to put those soils in jeopardy. The removal of the forest, the increase in destructive floods, the soil compaction and close-cropping wrought by grazing animals, plowing--all served to increase erosion".

Karl Marx wrote of the role of capitalism in soil depletion. In *Capital, Volume I*, he wrote:

> "All progress in capitalistic agriculture is a progress in the art, not only of robbing the labourer, but of robbing the soil; all progress in increasing the fertility of the soil for a given time, is a progress towards ruining the lasting sources of that fertility. The more a country starts its development on the foundation of modern industry, like the United States, for example, the more rapid is this process of destruction. Capitalist production, therefore, develops technology, and the combining together of various processes into a social whole, only by sapping the original sources of all wealth — the soil and the labourer."

One of the most widespread occurrences of soil depletion as of 2008 is in tropical zones where nutrient content of soils is low. The combined effects of growing population densities, large-scale industrial logging, slash-and-burn agriculture and ranching, and other factors, have in some places depleted soils through rapid and almost total nutrient removal.

The depletion of soil has affected the state of plant life and crops in agriculture in many countries. In the middle east for example, many countries in that find it difficult to grow produce because of droughts, lack of soil, and lack of irrigation. "The Middle East" has three countries that indicate a decline in crop production. the highest rates of productivity decline are found in hilly and dryland areas. Many countries in Africa also undergo a depletion of fertile soil. In regions of dry climate like Sudan and the countries that make up the Sahara Desert, droughts and soil degradation is common. Cash crops such as teas, maize, and beans that require a variety of nutrients in order to grow healthy. Soil fertility has decline in the farming regions of Africa and the use of artificial and natural fertilizers has been used to regain the nutrients of ground soil.

Topsoil depletion occurs when the nutrient-rich organic topsoil, which takes hundreds to thousands of years to build up under natural conditions, is eroded or depleted of its original organic material. Historically, many past civilizations' collapses can be attributed to the depletion of the topsoil. Since the beginning of agricultural production in the Great Plains of North America in the 1880s, about one-half of its topsoil has disappeared.

Depletion may occur through a variety of other effects, including overtillage (which damages soil structure), underuse of nutrient inputs which leads to mining of the soil nutrient bank, and salinization of soil.

Irrigation Water Effects

The quality of irrigation water is very important to maintain soil fertility and tilth, and for using more soil depth by the plants. When soil is irrigated with high alkaline water, unwanted sodium salts build up in the soil which would make soil draining capacity very poor. So plant roots can not penetrate deep into the soil for optimum growth in Alkali soils. When soil is irrigated with low pH / acidic water, the useful salts (Ca, Mg, K, P, S, etc.) are removed by draining water from the acidic soil and in addition unwanted aluminium and manganese salts to the plants are dissolved from the soil impeding plant growth. When soil is irrigated with high salinity water or sufficient water is not draining out from the irrigated soil, the soil would convert into saline soil or lose its fertility. Saline water enhance the turgor pressure or osmotic pressure requirement which impedes the off take of water and nutrients by the plant roots.

Top soil loss takes place in alkali soils due to erosion by rain water surface flows or drainage as they form colloids (fine mud) in contact with water. Plants absorb water-soluble inorganic salts only from the soil for their growth. Soil as such does not lose fertility just by growing crops but it lose its fertility due to accumulation of unwanted and depletion of wanted inorganic salts from the soil by improper irrigation and acid rain water (quantity and quality of water). The fertility of many soils which are not suitable for plant growth can be enhanced many times gradually by providing adequate irrigation water of suitable quality and good drainage from the soil.

AGROGEOLOGY

Agrogeology is the study of the origins of minerals known as agrominerals and their applications. These minerals are of importance to farming and horticulture, especially with regard to soil fertility and fertilizer components. These minerals are usually essential plant nutrients. Agrogeology can also be defined as the application of geology to problems in agriculture, particularly in reference to soil productivity and health. This field is a combination of a few different fields, including geology, soil science, agronomy, and chemistry. The overall objective is to advance agricultural production by using geological resources to improve chemical and physical aspects of soil.

Rock Phosphate as Fertilizer

A common problem faced in agriculture is dealing with soils lacking in phosphorus. Phosphorus, along with nitrogen and potassium, is an important element in determining plant development and health. A high percentage of traditional fertilizers intended to mend phosphorus-deficient soils end up becoming insoluble complexes in the soil. This presents a need for constant reapplication. Rock phosphate, also known as phosphorite, can be used as a sustainable, cost effective method to mend problems associated with plant growth.

Rock phosphate is mined from clay deposits that contain phosphorus. It can be found on across South Africa, Canada, sea beds, and sea mounts in the Pacific and Atlantic oceans. These rocks are mostly sedimentary, one example being limestone.

Unlike other elements that are soluble and easily accessible, rock phosphate needs to be processed in order to make the phosphorus in them available for plant and soil intake. Currently, there are a few ways of processing rock phosphate. Microbial solubilization of rock phosphate through fungi has been found to be able to break down inorganic phosphate into soluble forms by processes that produce organic acids.

Residual dust from mining has also been used in conjunction with processed fertilizer in order to improve plant development. A study in Zimbabwe suggests that this mixture increases plant growth, phosphorus levels, and organic carbon.

Commercial fertilizers mine and process rock phosphate using chemistry. Phosphorite is mined primarily by surface methods using draglines and bucket wheel excavators. Once it is ground and impurities are removed, water and sulfuric acid is added to the phosphate rock which generates gypsum crystals, a way of getting rid of what we don't want, leaving phosphorus as an acidic liquid. To raise phosphorus levels, impurities are precipitated out and any excess water is evaporated. Then vapor ammonia is applied to the liquid phosphorus and the end products are phosphorus granules.

Other Raw Materials Used in Agriculture

- Apatite: A major source of slow release of phosphate in acidic soils.

- Carbonate: Contains liming materials used to solve problems of acidity and related toxicities.

- Malachite: Useful for correction copper deficiencies.

- Scoria: Useful as a mulching material to conserve soil water and provide slow release of nutrients.

- Zeolite: Useful in conserving nitrogen and releasing phosphorus from apatite couple reaction, also raises pH.

FERTILIZERS

Fertilizers are used daily by farmers and families to help crops and gardens grow. Whether for a small garden of flowers and plants, or a large farm with thousands of acres of crops, a wide range of fertilizers have been developed to help different crops grow in different soil and weather conditions.

Chemical ingredients help create fertilizers that promote plant growth and are cost effective, too. Commercial and consumer fertilizers are strictly regulated by both individual states and the federal government to ensure that they are safe for the people who use them, people nearby, and the surrounding environment.

Uses and Benefits

With the global population steadily growing, it is important that enough crops are produced each year to provide food, clothing and other agricultural products to people around the world. Crops such as corn, wheat and cotton receive nutrients from the soil they are grown in; various crops deplete soil nutrients in different ways and rates. Some crop growth can deplete soil nutrients after just a few seasons of planting. Fertilizers play an important role in providing crops with the nutrients they need to grow and be harvested for nutritious food.

Fertilizers help deliver enough food to feed the world's population. But they can do even more. A class of fertilizers called micronutrient fertilizers is engineered to enrich crops with vital nutrients that help support human health. For example, micronutrients such as zinc are important to human nutrition, especially children. According to a United Nations study, much of the world's grain crops are grown in soil without adequate zinc; offering micronutrient fertilizers to grain crops enriches the grain with an important nutrient.

Safety Information

Fertilizers are essential to the security of the world's food supply, and they must be used properly. The manufacture, sale and transportation of fertilizers is heavily regulated. States have difference regulations and statutes that address fertilizer use and production to protect human health and the environment.

In addition to individual state regulations, the federal government also regulates fertilizer production, use and transport. Fertilizer producers are required to participate in the U.S. Occupational Safety and Health Administration's (OSHA) Process Safety Management Regulation to protect worker safety. In addition to OSHA, the U.S. Department of Agriculture, Environmental Protection Agency, Department of Homeland Security and Department of Transportation all play roles in regulating fertilizers and their production.

Ammonium nitrate, a good source of nitrogen and ammonium for plants, is an important ingredient in the production of high quality, effective fertilizers. Ammonium nitrate is produced using anhydrous ammonia, a gas that can become explosive when mixed with air. While this fertilizer is applied safely every day, this feature also means that it can be dangerous in the wrong hands. Fertilizer producers take great care in the safe use and storage of anhydrous ammonia and have implemented safety measures to help prevent or mitigate an incident.

Crops require a balanced diet of essential nutrients throughout their growth cycle.

Many of these essential nutrients can be found in the soil, but often in insufficient quantities to sustain high crop yields. Soil and climatic conditions can also limit a plant's uptake of nutrients at key growth stages.

Plants need 13 essential minerals, all of which play a number of important functions. If any of these is lacking, plant growth and yield suffer.

Crop Needs

Each crop needs a different range of nutrients at every critical stage of its development.

For example, nitrogen and phosphorous are often more critical at early stages of growth to fuel root and leaf development, whereas zinc and boron are important during flowering.

Cereal crops use nutrients for growth, progressively moving them from the roots, leaves and stems into the ear prior to the dying off and harvesting of the grain.

Tree crops have different nutrient requirements than field crops. They can store nutrients like nitrogen within their trunk, branches and leaves and then redistribute them at key points during the growth cycle. It is important, however, to supply trees with replacement levels of the nutrients removed in the harvested fruit and those that are critical for growth but can't be recycled.

High-value, high-quality greenhouse crops have perhaps the greatest need for nutritional precision. Top- quality strawberries, lettuce or fruit require a constant and accurately balanced diet. Growers therefore often control crop growth by spoon-feeding plants with what they need in an environment protected from the changing soil and weather conditions. In all situations, it is important that fertilizer formulation and selection matches the crop's needs.

Armed with this information and the right product, the grower can ensure nutrient needs are met and growth and crop quality is maximized. This provides high yields and high profits from the use of high-quality fertilizers.

Major Nutrients

Of the major nutrients, nitrogen (N) is often required in the greatest quantity by crops, primarily for vigor and yield. Nitrogen plays a key role in chlorophyll production and protein synthesis. Chlorophyll is the green plant pigment responsible for photosynthesis. When nitrogen is deficient, plants develop yellow or pale leaves and their growth is stunted.

Phosphorus (P), is a vital component of adenosine triphosphate (ATP) which supplies the energy for many processes in the plant. Phosphorus rarely produces spectacular growth responses, but is fundamental to the successful development of all crops. For example, maize or other corn crops that lack phosphorus during the growing season achieve lower yields.

Potassium (K) is needed by virtually all crops and often in higher rates than nitrogen. Potassium regulates the plant's water content and expansion. It is key to achieving good yield and quality in cotton and critical for increasing the size, juice content and sweetness of fruit.

Secondary Nutrients

Of the three secondary nutrients needed at lower levels than nitrogen, phosphorus and potassium (NPK), calcium (Ca) is perhaps the most important. Calcium strengthens cell walls, helping to reduce bruising and disease in fruit, salad and vegetable crops. This means that a good supply of calcium produces food crops that are less prone to damage and have a longer shelf life. Crops short in calcium will have growth disorders such as corky skin.

Fruit and vegetables containing higher levels of calcium also have a higher nutritional value – for example, vitamin C and antioxidants in tomatoes. This means that eating fresh fruit with strong skins and a great, crisp bite will help provide us with the calcium we need for strong bones.

Magnesium (Mg) is also important for crop quality, but is also a key component of leaf chlorophyll and the enzymes that support plant growth. Low magnesium leads to reduced photosynthesis, which severely limits crop yields. Grain fill in rice and dry matter content of potatoes can be significantly reduced if magnesium is undersupplied.

Sulfur (S) is an essential part of many amino acids and proteins. Without both S and Mg, crops suffer; growth slows and leaves turn pale or yellow. Sulfur is particularly important for ensuring the protein content of cereal crop grains.

Micronutrients

Micronutrients reinforce and supplement the strong plant growth and structures provided by major and secondary nutrients.

Most micronutrients influence growth. For example, manganese (Mn), iron (Fe) and copper (Cu) all influence photosynthesis, the process whereby plants use sunlight for growth.

- Iron deficiencies are common – for example in seed fruits – where the effect is to reduce production of chlorophyll. As a result, crops struggle and younger leaves develop a severe yellowing or chlorosis.

- Boron (B) is needed for the development of shoots and roots, and is essential during the flowering and fruiting phases of crops.

- Zinc (Zn) is needed for the production of important plant hormones, like auxin. Zinc deficiency leads to structural defects in leaves and other plant organs.

- Molybdenum (Mo) is involved in plant enzyme systems that control nitrogen metabolism.

MANURE

Manure is organic matter, mostly derived from animal feces except in the case of green manure, which can be used as organic fertilizer in agriculture. Manures contribute to the fertility of the soil by adding organic matter and nutrients, such as nitrogen, that are utilised by bacteria, fungi and other organisms in the soil. Higher organisms then feed on the fungi and bacteria in a chain of life that comprises the soil food web.

In the past, the term "manure" included inorganic fertilizers, but this usage is now very rare.

Types

There are in the 21st century three main classes of manures used in soil management:

Animal Manure

Concrete reservoirs, one new, and one containing cow manure mixed with water. This is common in rural Hainan Province, China.

Most animal manure consists of feces. Common forms of animal manure include farmyard manure (FYM) or farm slurry (liquid manure). FYM also contains plant material (often straw), which has been used as bedding for animals and has absorbed the feces and urine. Agricultural manure in liquid form, known as slurry, is produced by more intensive livestock rearing systems where concrete or slats are used, instead of straw bedding. Manure from different animals has different qualities and requires different application rates when used as fertilizer. For example horses, cattle, pigs, sheep, chickens, turkeys, rabbits, and guano from seabirds and bats all have different properties. For instance, sheep manure is high in nitrogen and potash, while pig manure is relatively low in both. Horses mainly eat grass and a few weeds so horse manure can contain grass and weed seeds, as horses do not digest seeds the way that cattle do. Cattle manure is a good source of nitrogen as well as organic carbon. Chicken litter, coming from a bird, is very concentrated in nitrogen and phosphate and is prized for both properties.

Animal manures may be adulterated or contaminated with other animal products, such as wool (shoddy and other hair), feathers, blood, and bone. Livestock feed can be mixed with the manure due to spillage. For example, chickens are often fed meat and bone meal, an animal product, which can end up becoming mixed with chicken litter.

Human Manure

Some people refer to human excreta as human manure, and the word "humanure" has also been used. Just like animal manure, it can be applied as a soil conditioner (reuse of excreta in agriculture). Sewage sludge is a material that contains human excreta, as it is generated after mixing excreta with water and treatment of the wastewater in a sewage treatment plant.

Compost

Compost containing turkey manure and wood chips from bedding material is
dried and then applied to pastures for fertilizer.

Compost is the decomposed remnants of organic materials. It is usually of plant origin, but often
includes some animal dung or bedding.

Green Manure

Green manures are crops grown for the express purpose of plowing them in, thus increasing fertility through the incorporation of nutrients and organic matter into the soil. Leguminous plants such as clover are often used for this, as they fix nitrogen using *Rhizobia* bacteria in specialized nodes in the root structure.

Other types of plant matter used as manure include the contents of the rumens of slaughtered ruminants, spent grain (left over from brewing beer) and seaweed.

Uses of Manure

Animal Manure

Manure on a wall.

Animal manure, such as chicken manure and cow dung, has been used for centuries as a fertilizer for farming. It can improve the soil structure (aggregation) so that the soil holds more nutrients and water, and therefore becomes more fertile. Animal manure also encourages soil microbial activity which promotes the soil's trace mineral supply, improving plant nutrition. It also contains some nitrogen and other nutrients that assist the growth of plants.

Manures with a particularly unpleasant odor (such as slurries from intensive pig farming) are usually knifed (injected) directly into the soil to reduce release of the odor. Manure from pigs and cattle is usually spread on fields using a manure spreader. Due to the relatively lower level of proteins in vegetable matter, herbivore manure has a milder smell than the dung of carnivores or omnivores. However, herbivore slurry that has undergone anaerobic fermentation may develop more unpleasant odors, and this can be a problem in some agricultural regions. Poultry droppings are harmful to plants when fresh, but after a period of composting are valuable fertilizers.

Manure is also commercially composted and bagged and sold as a soil amendment.

In 2018, Austrian scientists offered a method of paper production from elephant and cow manure.

Issues

The women of a neighborhood ward with manure on their way to the field of one of them.

Any quantity of manure may be a source of pathogens or food spoilage organisms which may be carried by flies, rodents or a range of other vector organisms and cause disease or put food safety at risk.

Livestock Antibiotics

In 2007, a University of Minnesota study indicated that foods such as corn, lettuce, and potatoes have been found to accumulate antibiotics from soils spread with animal manure that contains these drugs.

Organic foods may be much more or much less likely to contain antibiotics, depending on their sources and treatment of manure. For instance, by Soil Association Standard 4.7.38, most organic arable farmers either have their own supply of manure (which would, therefore, not normally contain drug residues) or else rely on green manure crops for the extra fertility (if any nonorganic manure is used by organic farmers, then it usually has to be rotted or composted to degrade any residues of drugs and eliminate any pathogenic bacteria—Standard 4.7.38, Soil Association organic farming standards). On the other hand, as found in the University of Minnesota study, the non-usage of artificial fertilizers, and resulting exclusive use of manure as fertilizer, by organic farmers can result in significantly greater accumulations of antibiotics in organic foods.

CONTOUR PLOUGHING

Contour bunding or contour farming or Contour ploughing is the farming practice of plowing and/or planting across a slope following its elevation contour lines. These contour lines create a water break which reduces the formation of rills and gullies during times of heavy water run-off; which is a major cause of soil erosion. The water break also allows more time for the water to settle into the soil. In contour plowing, the ruts made by the plow run perpendicular rather than parallel to the slopes, generally resulting in furrows that curve around the land and are level. This method is also known for preventing tillage erosion. Tillage erosion is the soil movement and erosion by tilling a given plot of land. A similar practice is contour bunding where stones are placed around the contours of slopes.Contour ploughing helps to reduce soil erosion.

Soil erosion prevention practices such as this can drastically decrease negative effects associated with soil erosion such as reduced crop productivity, worsened water quality, lower effective reservoir water levels, flooding, and habitat destruction. Contour farming is considered an active form of sustainable agriculture.

This was one of the main procedures promoted by the UK Soil Conservation Service (the current Natural Resources Conservation Service) during the 1930s. The US Department of Agriculture established the Soil Conservation Service in 1935 during the Dust Bowl when it became apparent that soil erosion was a huge problem along with desertification.

The extent of the problem was such that the 1934 "Yearbook of Agriculture" noted that Approximately 35 million acres [142,000 km²] of formerly cultivated land have essentially been destroyed for crop production. 100 million acres [405,000 km²] now in crops have lost all or most of the topsoil; 125 million acres [506,000 km²] of land now in crops are rapidly losing topsoil. This can lead to large scale desertification which can permanently transform a formerly productive landscape to an arid one that becomes increasingly intensive and expensive to farm.

The Soil Conservation Service worked with state governments and universities with established agriculture programs such as the University of Nebraska to promote the method to farmers. By 1938, the introduction of new agricultural techniques such as contour plowing had reduced the loss of soil by 65% despite the continuation of the drought.

Demonstrations showed that contour farming, under ideal conditions, will increase yields of row crops by up to 50%, with increases of between 5 and 10% being common. Importantly, the technique also significantly reduces soil erosion, fertilizer loss, and overall makes farming less energy and resource intensive under most circumstances. Reducing fertilizer loss not only saves the farmer time and money, but it also decreases risk of harming regional freshwater systems. Soil erosion caused from heavy rain can encourage the development of rills and gullies which carry excess nutrients into freshwater systems through the process of eutrophication.

Contour plowing is also promoted in countries with similar rainfall patterns to the United States such as western Canada and Australia.

The practice is effective only on slopes with between 2% and 10% gradient and when rainfall does not exceed a certain amount within a certain period. On steeper slopes and areas with greater rainfall, a procedure known as strip cropping is used with contour farming to provide additional protection. Contour farming is most effective when used with other soil conservation methods like strip cropping, terrace (agriculture) farming, and the use of a cover crop. The proper combination of such farming methods can be determined by various climatic and soil conditions of that given area. Farming sites are often classified into five levels: insensitive, mild, moderate, high and extreme, depending on the regions soil sensitivity. Contour farming is applied in certain European countries such as Belgium, Italy, Greece, Romania, Slovenia and Spain in areas with higher than 10% slope.

P. A. Yeomans' Keyline Design system is critical of traditional contour plowing techniques, and improves the system through observing normal land form and topography. At one end of a contour the slope of the land will always be steeper than at the other. Thus when plowing parallel runs paralleling any contour the plow furrows soon deviate from a true contour. Rain water in these furrows will thus flow sideways along the falling "contour" line. This can often concentrate water in a ways that exacerbates erosion instead of reducing it. Yeomans was the first to appreciate the significance of this phenomenon. Keyline cultivation utilizes this "off contour" drift in cultivating furrows to control the movement of rain water for the benefit of the land.

Contour bunding has been widely adopted in Burkina Faso after it was suggested by British Oxfam worker Bill Hereford in the beginning of the 1980s.

AEROPONICS

Aeroponics is the process of growing plants in an air or mist environment without the use of soil or an aggregate medium (known as geoponics). Aeroponic culture differs from both conventional hydroponics, aquaponics, and in-vitro (plant tissue culture) growing. Unlike hydroponics, which uses a liquid nutrient solution as a growing medium and essential minerals to sustain plant growth; or aquaponics which uses water and fish waste, aeroponics is conducted without a growing medium.It is sometimes considered a type of hydroponics, since water is used in aeroponics to transmit nutrients.

Methods

The basic principle of aeroponic growing is to grow plants suspended in a closed or semi-closed environment by spraying the plant's dangling roots and lower stem with an atomized or sprayed, nutrient-rich water solution. The leaves and crown, often called the canopy, extend above. The roots of the plant are separated by the plant support structure. Often, closed-cell foam is compressed around the lower stem and inserted into an opening in the aeroponic chamber, which decreases labor and expense; for larger plants, trellising is used to suspend the weight of vegetation and fruit.

Ideally, the environment is kept free from pests and disease so that the plants may grow healthier and more quickly than plants grown in a medium. However, since most aeroponic environments are not perfectly closed off to the outside, pests and disease may still cause a threat. Controlled environments advance plant development, health, growth, flowering and fruiting for any given plant species and cultivars.

Due to the sensitivity of root systems, aeroponics is often combined with conventional hydroponics, which is used as an emergency "crop saver" – backup nutrition and water supply – if the aeroponic apparatus fails.

High-pressure aeroponics is defined as delivering nutrients to the roots via 20–50 micrometre mist heads using a high-pressure (80 pounds per square inch (550 kPa)) diaphragm pump.

Increased Air Exposure

Close-up of the first patented aeroponic plant support structure. Its unrestricted support of the plant allows for normal growth in the air/moisture environment, and is still in use today.

Air cultures optimize access to air for successful plant growth. Materials and devices which hold and support the aeroponic grown plants must be devoid of disease or pathogens. A distinction of a true aeroponic culture and apparatus is that it provides plant support features that are minimal. Minimal contact between a plant and support structure allows for 100% of the plant to be entirely in air. Long-term aeroponic cultivation requires the root systems to be free of constraints surrounding the stem and root systems. Physical contact is minimized so that it does not hinder natural growth and root expansion or access to pure water, air exchange and disease-free conditions.

Benefits of Oxygen in the Root Zone

Oxygen (O_2) in the rhizosphere (root zone) is necessary for healthy plant growth. As aeroponics is conducted in air combined with micro-droplets of water, almost any plant can grow to maturity in air with a plentiful supply of oxygen, water and nutrients.

Some growers favor aeroponic systems over other methods of hydroponics because the increased aeration of nutrient solution delivers more oxygen to plant roots, stimulating growth and helping to prevent pathogen formation.

Clean air supplies oxygen which is an excellent purifier for plants and the aeroponic environment. For natural growth to occur, the plant must have unrestricted access to air. Plants must be allowed to grow in a natural manner for successful physiological development. The more confining the plant support becomes, the greater incidence of increasing disease pressure of the plant and the aeroponic system.

Some researchers have used aeroponics to study the effects of root zone gas composition on plant performance. Soffer and Burger studied the effects of dissolved oxygen concentrations on the formation of adventitious roots in what they termed "aero-hydroponics." They utilized a 3-tier hydro and aero system, in which three separate zones were formed within the root area. The ends of the roots were submerged in the nutrient reservoir, while the middle of the root section received nutrient mist and the upper portion was above the mist. Their results showed that dissolved O_2 is essential to root formation, but went on to show that for the three O_2 concentrations tested, the number of roots and root length were always greater in the central misted section than either the submersed section or the un-misted section. Even at the lowest concentration, the misted section rooted successfully.

Other Benefits of Air (CO_2)

Plants in a true aeroponic apparatus have 100% access to the CO_2 concentrations ranging from 450 ppm to 780 ppm for photosynthesis. At one mile (1.6 km) above sea level, the CO_2 concentration in the air is 450 ppm during daylight. At night, the CO_2 level will rise to 780 ppm. Lower elevations will have higher levels. In any case, the air culture apparatus offers the ability for plants to have full access to all of the available CO_2 in the air for photosynthesis.

Growing under lights during the evening allows aeroponics to benefit from the natural occurrence.

Disease-free Cultivation

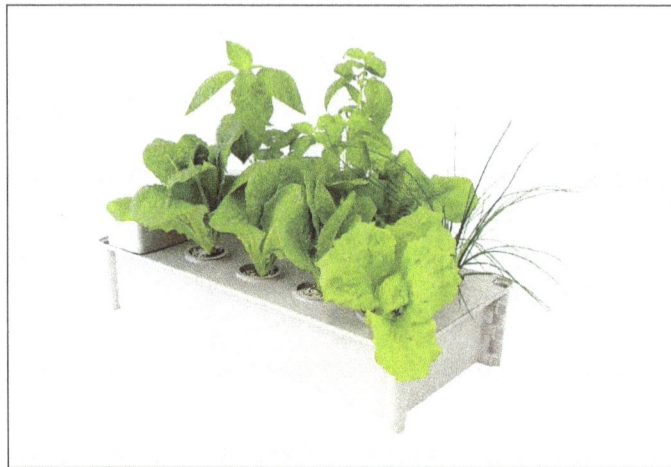

Basil grown from seed in an aeroponic system located inside
a modern greenhouse was first achieved 1986.

Aeroponics can limit disease transmission since plant-to-plant contact is reduced and each spray pulse can be sterile. In the case of soil, aggregate, or other media, disease can spread throughout the growth media, infecting many plants. In most greenhouses, these solid media require sterilization after each crop and, in many cases, they are simply discarded and replaced with fresh, sterile media.

A distinct advantage of aeroponic technology is that if a particular plant does become diseased, it can be quickly removed from the plant support structure without disrupting or infecting the other plants.

Due to the disease-free environment that is unique to aeroponics, many plants can grow at higher density (plants per square meter) when compared to more traditional forms of cultivation (hydroponics, soil and Nutrient Film Technique [NFT]). Commercial aeroponic systems incorporate hardware features that accommodate the crop's expanding root systems.

Researchers have described aeroponics as a "valuable, simple, and rapid method for preliminary screening of genotypes for resistance to specific seedling blight or root rot."

The isolating nature of the aeroponic system allowed them to avoid the complications encountered when studying these infections in soil culture.

Water and Nutrient Hydro-atomization

Aeroponic equipment involves the use of sprayers, misters, foggers, or other devices to create a fine mist of solution to deliver nutrients to plant roots. Aeroponic systems are normally closed-looped systems providing macro and micro-environments suitable to sustain a reliable, constant air culture. Numerous inventions have been developed to facilitate aeroponic spraying and misting. The key to root development in an aeroponic environment is the size of the water droplet. In commercial applications, a hydro-atomizing spray at 360° is employed to cover large areas of roots utilizing air pressure misting.

A variation of the mist technique employs the use of ultrasonic foggers to mist nutrient solutions in low pressure aeroponic devices.

Water droplet size is crucial for sustaining aeroponic growth. Too large a water droplet means less oxygen is available to the root system. Too fine a water droplet, such as those generated by the ultrasonic mister, produce excessive root hair without developing a lateral root system for sustained growth in an aeroponic system.

Mineralization of the ultrasonic transducers requires maintenance and potential for component failure. This is also a shortcoming of metal spray jets and misters. Restricted access to the water causes the plant to lose turgidity and wilt.

Advanced Materials

NASA has funded research and development of new advanced materials to improve aeroponic reliability and maintenance reduction. It also has determined that high pressure hydro-atomized mist of 5-50 micrometres micro-droplets is necessary for long-term aeroponic growing.

For long-term growing, the mist system must have significant pressure to force the mist into the dense root system(s). Repeatability is the key to aeroponics and includes the hydro-atomized droplet size. Degradation of the spray due to mineralization of mist heads inhibits the delivery of the water nutrient solution, leading to an environmental imbalance in the air culture environment.

Special low-mass polymer materials were developed and are used to eliminate mineralization in next generation hydro-atomizing misting and spray jets.

Nutrient Uptake

The discrete nature of interval and duration aeroponics allows the measurement of nutrient uptake over time under varying conditions. Barak et al. used an aeroponic system for non-destructive measurement of water and ion uptake rates for cranberries.

Close-up of roots grown from wheat seed using aeroponics.

In their study, these researchers found that by measuring the concentrations and volumes of input and efflux solutions, they could accurately calculate the nutrient uptake rate (which was verified by comparing the results with N-isotope measurements). After verification of their analytical method, Barak et al. went on to generate additional data specific to the cranberry, such as diurnal variation in nutrient uptake, correlation between ammonium uptake and proton efflux, and the relationship between ion concentration and uptake. Work such as this not only shows the promise of aeroponics as a research tool for nutrient uptake, but also opens up possibilities for the monitoring of plant health and optimization of crops grown in closed environments.

Atomization (>65 pounds per square inch (450 kPa)), increases bioavailability of nutrients, consequently, nutrient strength must be significantly reduced or leaf and root burn will develop. This is caused by the feed cycle being too long or the pause cycle too short; either discourages both lateral root growth and root hair development. Plant growth and fruiting times are significantly shortened when feed cycles are as short as possible. Ideally, roots should never be more than slightly damp nor overly dry. A typical feed/pause cycle is < 2 seconds on, followed by ~1.5-2 minute pause-24/7, however, when an accumulator system is incorporated, cycle times can be further reduced to < ~1 second on, ~1 minute pause.

As a Research Tool

Soon after its development, aeroponics took hold as a valuable research tool. Aeroponics offered researchers a noninvasive way to examine roots under development. This new technology also allowed researchers a larger number and a wider range of experimental parameters to use in their work.

The ability to precisely control the root zone moisture levels and the amount of water delivered makes aeroponics ideally suited for the study of water stress. K. Hubick evaluated aeroponics as a means to produce consistent, minimally water-stressed plants for use in drought or flood physiology experiments.

Aeroponics is the ideal tool for the study of root morphology. The absence of aggregates offers researchers easy access to the entire, intact root structure without the damage that can be caused by removal of roots from soils or aggregates. It's been noted that aeroponics produces more normal root systems than hydroponics.

Terminology

- Aeroponic growing refers to plants grown in an air culture that can develop and grow in a normal and natural manner.

- Aeroponic growth refers to growth achieved in an air culture.

- Aeroponic system refers to hardware and system components assembled to sustain plants in an air culture.

- Aeroponic greenhouse refers to a climate controlled glass or plastic structure with equipment to grow plants in air/mist environment.

- Aeroponic conditions refers to air culture environmental parameters for sustaining plant growth for a plant species.

- Aeroponic roots refers to a root system grown in an air culture.

Types of Aeroponics

Low-pressure Units

In most low-pressure aeroponic gardens, the plant roots are suspended above a reservoir of nutrient solution or inside a channel connected to a reservoir. A low-pressure pump delivers nutrient solution via jets or by ultrasonic transducers, which then drips or drains back into the reservoir. As plants grow to maturity in these units they tend to suffer from dry sections of the root systems, which prevent adequate nutrient uptake. These units, because of cost, lack features to purify the nutrient solution, and adequately remove incontinuities, debris, and unwanted pathogens. Such units are usually suitable for bench top growing and demonstrating the principles of aeroponics.

High-pressure Devices

High-pressure aeroponic techniques, where the mist is generated by high-pressure pump(s), are typically used in the cultivation of high value crops and plant specimens that can offset the high setup costs associated with this method of horticulture.

High-pressure aeroponics systems include technologies for air and water purification, nutrient sterilization, low-mass polymers and pressurized nutrient delivery systems.

Commercial Systems

Commercial aeroponic systems comprise high-pressure device hardware and biological systems. The biological systems matrix includes enhancements for extended plant life and crop maturation.

Biological subsystems and hardware components include effluent controls systems, disease prevention, pathogen resistance features, precision timing and nutrient solution pressurization, heating and cooling sensors, thermal control of solutions, efficient photon-flux light arrays, spectrum filtration spanning, fail-safe sensors and protection, reduced maintenance & labor saving features, and ergonomics and long-term reliability features.

Commercial aeroponic systems, like the high-pressure devices, are used for the cultivation of high value crops where multiple crop rotations are achieved on an ongoing commercial basis.

Advanced commercial systems include data gathering, monitoring, analytical feedback and internet connections to various subsystems.

MULCH

A mulch is a layer of material applied to the surface of soil. Reasons for applying mulch include conservation of soil moisture, improving fertility and health of the soil, reducing weed growth and enhancing the visual appeal of the area.

A mulch is usually, but not exclusively, organic in nature. It may be permanent (e.g. plastic sheeting) or temporary (e.g. bark chips). It may be applied to bare soil or around existing plants. Mulches of manure or compost will be incorporated naturally into the soil by the activity of worms and other organisms. The process is used both in commercial crop production and in gardening, and when applied correctly, can dramatically improve soil productivity.

Uses

Many materials are used as mulches, which are used to retain soil moisture, regulate soil temperature, suppress weed growth, and for aesthetics. They are applied to the soil surface, around trees, paths, flower beds, to prevent soil erosion on slopes, and in production areas for flower and vegetable crops. Mulch layers are normally 2 inches (5.1 cm) or more deep when applied.

They are applied at various times of the year depending on the purpose. Towards the beginning of the growing season, mulches serve initially to warm the soil by helping it retain heat which is lost during the night. This allows early seeding and transplanting of certain crops, and encourages faster growth. As the season progresses, mulch stabilizes the soil temperature and moisture, and prevents the growing of weeds from seeds. In temperate climates, the effect of mulch is dependent upon the time of year they are applied and when applied in fall and winter, are used to delay the growth of perennial plants in the spring or prevent growth in winter during warm spells, which limits freeze thaw damage.

The effect of mulch upon soil moisture content is complex. Mulch forms a layer between the soil and the atmosphere preventing sunlight from reaching the soil surface, thus reducing evaporation. However, mulch can also prevent water from reaching the soil by absorbing or blocking water from light rains.

In order to maximise the benefits of mulch, while minimizing its negative influences, it is often applied in late spring/early summer when soil temperatures have risen sufficiently, but soil moisture content is still relatively high. However, permanent mulch is also widely used and valued for its simplicity, as popularized by author Ruth Stout, who said, "My way is simply to keep a thick mulch of any vegetable matter that rots on both sides of my vegetable and flower garden all year long. As it decays and enriches the soils."

Plastic mulch used in large-scale commercial production is laid down with a tractor-drawn or standalone layer of plastic mulch. This is usually part of a sophisticated mechanical process, where raised beds are formed, plastic is rolled out on top, and seedlings are transplanted through it. Drip irrigation is often required, with drip tape laid under the plastic, as plastic mulch is impermeable to water.

Materials

Rubber mulch nuggets in a playground. The white fibers are nylon cords,
which are present in the tires from which the mulch is made.

Materials used as mulches vary and depend on a number of factors. Use takes into consideration availability, cost, appearance, the effect it has on the soil—including chemical reactions and pH, durability, combustibility, rate of decomposition, how clean it is—some can contain weed seeds or plant pathogens.

Crushed stone mulch

A variety of materials are used as mulch:

- Organic residues: grass clippings, leaves, hay, straw, kitchen scraps comfrey, shredded bark, whole bark nuggets, sawdust, shells, woodchips, shredded newspaper, cardboard, wool, animal manure, etc. Many of these materials also act as a direct composting system, such as the mulched clippings of a mulching lawn mower, or other organics applied as sheet composting.

- Compost: fully composted materials are used to avoid possible phytotoxicity problems. Materials that are free of seeds are ideally used, to prevent weeds being introduced by the mulch.

- Old carpet (synthetic or natural): makes a free, readily available mulch.

- Rubber mulch: made from recycled tire rubber.

- Plastic mulch: crops grow through slits or holes in thin plastic sheeting. This method is predominant in large-scale vegetable growing, with millions of acres cultivated under plastic mulch worldwide each year (disposal of plastic mulch is cited as an environmental problem).

- Rock and gravel can also be used as a mulch. In cooler climates the heat retained by rocks may extend the growing season.

In some areas of the United States, such as central Pennsylvania and northern California, mulch is often referred to as "tanbark", even by manufacturers and distributors. In these areas, the word "mulch" is used specifically to refer to very fine tanbark or peat moss.

Organic Mulches

Mulching coconut farm.

Organic mulches decay over time and are temporary. The way a particular organic mulch decomposes and reacts to wetting by rain and dew affects its usefulness.

Some mulches such as straw, peat, sawdust and other wood products may for a while negatively affect plant growth because of their wide carbon to nitrogen ratio, because bacteria and fungi that decompose the materials remove nitrogen from the surrounding soil for growth. However, whether this effect has any practical impact on gardens is disputed by researchers and the experience of gardeners. Organic mulches can mat down, forming a barrier that blocks water and air flow between the soil and the atmosphere. Vertically applied organic mulches can wick water from the soil to the surface, which can dry out the soil. Mulch made with wood can contain or feed termites, so care must be taken about not placing mulch too close to houses or building that can be damaged by those insects. Some mulch manufacturers recommend putting mulch several inches away from buildings.

Commonly available organic mulches include:

- Leaves from deciduous trees, which drop their foliage in the autumn/fall. They tend to be dry and blow around in the wind, so are often chopped or shredded before application. As they decompose they adhere to each other but also allow water and moisture to seep down to the soil surface. Thick layers of entire leaves, especially of maples and oaks, can form a soggy mat in winter and spring which can impede the new growth lawn grass and other plants. Dry leaves are used as winter mulches to protect plants from freezing and thawing in areas with cold winters; they are normally removed during spring.

- Grass clippings, from mowed lawns are sometimes collected and used elsewhere as mulch. Grass clippings are dense and tend to mat down, so are mixed with tree leaves or rough compost to provide aeration and to facilitate their decomposition without smelly putrefaction. Rotting fresh grass clippings can damage plants; their rotting often produces a damaging buildup of trapped heat. Grass clippings are often dried thoroughly before application, which mediates against rapid decomposition and excessive heat generation. Fresh green grass clippings are relatively high in nitrate content, and when used as a mulch, much of the nitrate is returned to the soil, conversely the routine removal of grass clippings from the lawn results in nitrogen deficiency for the lawn.

- Peat moss, or sphagnum peat, is long lasting and packaged, making it convenient and popular as a mulch. When wetted and dried, it can form a dense crust that does not allow water to soak in. When dry it can also burn, producing a smoldering fire. It is sometimes mixed with pine needles to produce a mulch that is friable. It can also lower the pH of the soil surface, making it useful as a mulch under acid loving plants.

  However peat bogs are a valuable wildlife habitat, and peat is also one of the largest stores of carbon (in Britain, out of a total estimated 9952 million tonnes of carbon in British vegetation and soils, 6948 million tonnes carbon are estimated to be in Scottish, mostly peatland, soils), so gardeners who wish to protect the environment will choose more sustainable alternatives.

- Wood chips are a byproduct of the pruning of trees by arborists, utilities and parks; they are used to dispose of bulky waste. Tree branches and large stems are rather coarse after chipping and tend to be used as a mulch at least three inches thick. The chips are used to conserve soil moisture, moderate soil temperature and suppress weed growth. The decay of freshly produced chips from recently living woody plants, consumes nitrate; this is often off set with a light application of a high-nitrate fertilizer. Wood chips are most often used under trees and shrubs. When used around soft stemmed plants, an unmulched zone is left around the plant stems to prevent stem rot or other possible diseases. They are often used to mulch trails, because they are readily produced with little additional cost outside of the normal disposal cost of tree maintenance. Wood chips come in various colors.

- Woodchip mulch is a byproduct of reprocessing used (untreated) timber (usually packaging pallets), to dispose of wood waste by creating woodchip mulch. The chips are used to conserve soil moisture, moderate soil temperature and suppress weed growth. Woodchip mulch is often used under trees, shrubs or large planting areas and can last much longer than arborist mulch. In addition, many consider woodchip mulch to be visually appealing, as it comes in various colors. Woodchips can also be reprocessed into playground woodchip to be used as an impact-attenuating playground surfacing.

Bark chips.

- Bark chips of various grades are produced from the outer corky bark layer of timber trees. Sizes vary from thin shredded strands to large coarse blocks. The finer types are very attractive but have a large exposed surface area that leads to quicker decay. Layers two or three inches deep are usually used, bark is relativity inert and its decay does not demand soil nitrates. Bark chips are also available in various colors.

- Straw mulch or field hay or salt hay are lightweight and normally sold in compressed bales. They have an unkempt look and are used in vegetable gardens and as a winter covering. They are biodegradable and neutral in pH. They have good moisture retention and weed controlling properties but also are more likely to be contaminated with weed seeds. Salt hay is less likely to have weed seeds than field hay. Straw mulch is also available in various colors.

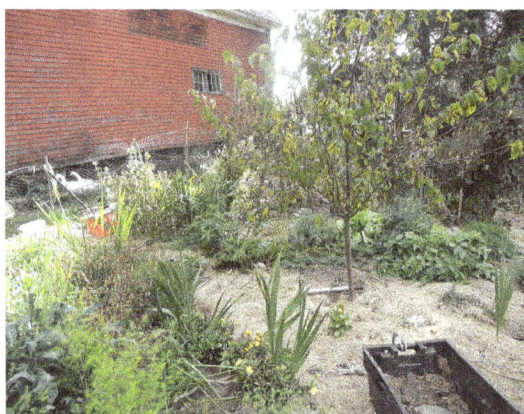

Permaculture garden with a fruit tree, herbs, flowers and vegetables mulched with hay.

- Cardboard or newspaper can be used as mulches. These are best used as a base layer upon which a heavier mulch such as compost is placed to prevent the lighter cardboard/

newspaper layer from blowing away. By incorporating a layer of cardboard/newspaper into a mulch, the quantity of heavier mulch can be reduced, whilst improving the weed suppressant and moisture retaining properties of the mulch. However, additional labour is expended when planting through a mulch containing a cardboard/newspaper layer, as holes must be cut for each plant. Sowing seed through mulches containing a cardboard/newspaper layer is impractical. Application of newspaper mulch in windy weather can be facilitated by briefly pre-soaking the newspaper in water to increase its weight.

- Synthetic carpet that is composed of artificial fibers may be removed after planting to prevent fibers taking a long time to decompose, whereas carpet made from natural fibers may be kept in place, blocking competition from weeds. Rain is absorbed by carpet and then slowly released into the soil, reducing watering needs.

Colored Mulch

Some organic mulches are colored red, brown, black, and other colors. Isopropanolamine, specifically 1-Amino-2-propanol or DOW™ monoisopropanolamine, may be used as a pigment dispersant and color fastener in these mulches. Types of mulch which can be dyed include: wood chips, bark chips (barkdust) and pine straw. Colored mulch is made by dyeing the mulch in a water-based solution of colorant and chemical binder. When colored mulch first entered the market, most formulas were suspected to contain toxic, heavy metals and other contaminates. Today, "current investigations indicate that mulch colorants pose no threat to people, pets or the environment. The dyes currently used by the mulch and soil industry are similar to those used in the cosmetic and other manufacturing industries (i.e., iron oxide)," as stated by the Mulch and Soil Council. Colored mulch can be applied anywhere non-colored mulch is used (such as large bedded areas or around plants) and features many of the same gardening benefits as traditional mulch, such as improving soil productivity and retaining moisture. As mulch decomposes, just as with non-colored mulch, more mulch may need to be added to continue providing benefits to the soil and plants. However, if mulch is faded, spraying dye to previously spread mulch in order to restore color is an option.

Anaerobic (Sour) Mulch

Mulch normally smells like freshly cut wood, but sometimes develops a toxicity that causes it to smell like vinegar, ammonia, sulfur or silage. This happens when material with ample nitrogen content is not rotated often enough and it forms pockets of increased decomposition. When this occurs, the process may become anaerobic and produce these phytotoxic materials in small quantities. Once exposed to the air, the process quickly reverts to an aerobic process, but these toxic materials may be present for a period of time. If the mulch is placed around plants before the toxicity has had a chance to dissipate, then the plants could very likely be damaged or killed depending on their hardiness. Plants that are predominantly low to the ground or freshly planted are the most susceptible, and the phytotoxicity may prevent germination of some seeds.

If sour mulch is applied and there is plant kill, the best thing to do is to water the mulch heavily. Water dissipates the chemicals faster and refreshes the plants. Removing the offending mulch may have little effect, because by the time plant kill is noticed, most of the toxicity is already dissipated. While testing after plant kill will not likely turn up anything, a simple pH check may reveal high acidity, in the range of 3.8 to 5.6 instead of the normal range of 6.0 to 7.2. Finally, placing a bit of the offending mulch around another plant to check for plant kill will verify if the toxicity has departed. If the new plant is also killed, then sour mulch is probably not the problem.

Groundcovers (living mulches)

Groundcovers are plants which grow close to the ground, under the main crop, to slow the development of weeds and provide other benefits of mulch. They are usually fast-growing plants that continue growing with the main crops. By contrast, cover crops are incorporated into the soil or killed with herbicides. However, live mulches also may need to be mechanically or chemically killed eventually to prevent competition with the main crop.

Some groundcovers can perform additional roles in the garden such as nitrogen fixation in the case of clovers, dynamic accumulation of nutrients from the subsoil in the case of creeping comfrey (*Symphytum ibericum*), and even food production in the case of *Rubus tricolor*.

On-site Production

Owing to the great bulk of mulch which is often required on a site, it is often impractical and expensive to source and import sufficient mulch materials. An alternative to importing mulch materials is to grow them on site in a "mulch garden" – an area of the site dedicated entirely to the production of mulch which is then transferred to the growing area. Mulch gardens should be sited as close as possible to the growing area so as to facilitate transfer of mulch materials.

Mulching (composting) over Unwanted Plants

Sufficient mulch over plants will destroy them, and may be more advantageous than using herbicide, cutting, mowing, pulling, raking, or tilling. The higher the temperature that this "mulch" is composted, the quicker the reduction of undesirable materials. "Undesirable materials" may include living seed, plant "trash", as well as pathogens such as from animal feces, urine (e.g. hantavirus), fleas, lice, ticks, etc.

In some ways this improves the soil by attracting and feeding earthworms, and adding humus. Earthworms "till" the soil, and their feces are among the best fertilizers and soil conditioners.

Urine may be toxic to plants if applied to growing areas undiluted. See Compost ingredients: Human Waste.

Polypropylene and Polyethylene Mulch

Polypropylene mulch is made up of polypropylene polymers where polyethylene mulch is made up of polyethylene polymers. These mulches are commonly used in many plastics. Polyethylene is used mainly for weed reduction, where polypropylene is used mainly on perennials. This mulch is placed on top of the soil and can be done by machine or hand with pegs to keep the mulch tight against the soil. This mulch can prevent soil erosion, reduce weeding, conserve soil moisture, and increase temperature of the soil. Ultimately this can reduce the amount of work a farmer may have to do, and the amount of herbicides applied during the growing period. The black and clear mulches capture sunlight and warm the soil increasing the growth rate. White and other reflective colours will also warm the soil, but they do not suppress weeds as well. This mulch may require other sources of obtaining water such as drip irrigation since it can reduce the amount of water that reaches the soil. This mulch needs to be manually removed at the end of the season since when it starts to break down it breaks down into smaller pieces. If the mulch is not removed before it starts to break down eventually it will break down into ketones and aldehydes polluting the soil. This mulch is technically biodegradable but does not break down into the same materials the more natural biodegradable mulch does.

Biodegradable Mulch

Quality biodegradable mulches are made out of plant starches and sugars or polyester fibers. These starches can come from plants such as wheat and corn. These mulch films may be a bit more permeable allowing more water into the soil. This mulch can prevent soil erosion, reduce weeding, conserve soil moisture, and increase temperature of the soil. Ultimately this can reduce the amount of herbicides used and manual labor farmers may have to do throughout the growing season. At the end of the season these mulches will start to break down from heat. Microorganisms in the soil break down the mulch into two components, water and CO_2, leaving no toxic residues behind. This source of mulch is even less manual labor since it does not need to be removed at the end of the season and can actually be tilled into the soil. With this mulch it is important to take into consideration that it's much more delicate than other kinds. It should be placed on a day which is not too hot and with less tension than other synthetic mulches. These also can be placed by machine or hand but it is ideal to have a more starchy mulch that will allow it to stick to the soil better.

TILLAGE

Tillage is the agricultural preparation of soil by mechanical agitation of various types, such as digging, stirring, and overturning. Examples of human-powered tilling methods using hand tools include shovelling, picking, mattock work, hoeing, and raking. Examples of draft-animal-powered or mechanized work include ploughing (overturning with moldboards or chiseling with chisel shanks), rototilling, rolling with cultipackers or other rollers, harrowing, and cultivating with cultivator shanks (teeth). Small-scale gardening and farming, for household food production or small business production, tends to use the smaller-scale methods, whereas medium- to large-scale farming tends to use the larger-scale methods.

Tillage that is deeper and more thorough is classified as primary, and tillage that is shallower and sometimes more selective of location is secondary. Primary tillage such as ploughing tends to produce a rough surface finish, whereas secondary tillage tends to produce a smoother surface finish, such as that required to make a good seedbed for many crops. Harrowing and rototilling often combine primary and secondary tillage into one operation.

"Tillage" can also mean the land that is tilled. The word "cultivation" has several senses that overlap substantially with those of "tillage". In a general context, both can refer to agriculture. Within agriculture, both can refer to any kind of soil agitation. Additionally, "cultivation" or "cultivating" may refer to an even narrower sense of shallow, selective secondary tillage of row crop fields that kills weeds while sparing the crop plants.

Tilling was first performed via human labor, sometimes involving slaves. Hoofed animals could also be used to till soil by trampling. The wooden plow was then invented. It could be pulled with human labor, or by mule, ox, elephant, water buffalo, or similar sturdy animal. Horses are generally unsuitable, though breeds such as the Clydesdale were bred as draft animals. The steel plow allowed farming in the American Midwest, where tough prairie grasses and rocks caused trouble. Soon after 1900, the farm tractor was introduced, which eventually made modern large-scale agriculture possible.

Primary and Secondary Tillage

Primary tillage is usually conducted after the last harvest, when the soil is wet enough to allow plowing but also allows good traction. Some soil types can be plowed dry. The objective of primary tillage is to attain a reasonable depth of soft soil, incorporate crop residues, kill weeds, and to aerate the soil. Secondary tillage is any subsequent tillage, in order to incorporate fertilizers, reduce the soil to a finer tilth, level the surface, or control weeds.

Reduced Tillage

Reduced tillage leaves between 15 and 30% crop residue cover on the soil or 500 to 1000 pounds per acre (560 to 1100 kg/ha) of small grain residue during the critical erosion period. This may involve the use of a chisel plow, field cultivators, or other implements.

Plough tilling the field.

Intensive Tillage

Intensive tillage leaves less than 15% crop residue cover or less than 500 pounds per acre (560 kg/ha) of small grain residue. This type of tillage is often referred to as conventional tillage, but as conservational tillage is now more widely used than intensive tillage (in the United States), it is often not appropriate to refer to this type of tillage as conventional. Intensive tillage often involves multiple operations with implements such as a mold board, disk, and/or chisel plow. After this, a finisher with a harrow, rolling basket, and cutter can be used to prepare the seed bed. There are many variations.

Conservation Tillage

Conservation tillage leaves at least 30% of crop residue on the soil surface, or at least 1,000 lb/ac (1,100 kg/ha) of small grain residue on the surface during the critical soil erosion period. This slows water movement, which reduces the amount of soil erosion. Additionally, conservation tillage has been found to benefit predatory arthropods that can enhance pest control. Conservation tillage also benefits farmers by reducing fuel consumption and soil compaction. By reducing the number of times the farmer travels over the field, farmers realize significant savings in fuel and labor. In most years since 1997, conservation tillage was used in US cropland more than intensive or reduced tillage.

However, conservation tillage delays warming of the soil due to the reduction of dark earth exposure to the warmth of the spring sun, thus delaying the planting of the next year's spring crop of corn.

- No-till - Never use a plow, disk, etc. ever again. Aims for 100% ground cover.

- Strip-till - Narrow strips are tilled where seeds will be planted, leaving the soil in between the rows untilled.

- Mulch-till.

- Rotational Tillage - Tilling the soil every two years or less often (every other year, or every third year, etc).

- Ridge-till.

Zone Tillage

Zone tillage is a form of modified deep tillage in which only narrow strips are tilled, leaving soil in between the rows untilled. This type of tillage agitates the soil to help reduce soil compaction problems and to improve internal soil drainage. It is designed to only disrupt the soil in a narrow strip directly below the crop row. In comparison to no-till, which relies on the previous year's plant residue to protect the soil and aides in postponement of the warming of the soil and crop growth in Northern climates, zone tillage creates approximately a strip approximately five inches wide that simultaneously breaks up plow pans, assists in warming the soil and helps to prepare a seedbed. When combined with cover crops, zone tillage helps replace lost organic matter, slows the deterioration of the soil, improves soil drainage, increases soil water and nutrient holding capacity, and allows necessary soil organisms to survive.

It has been successfully used on farms in the Midwest and West for over 40 years, and is currently used on more than 36% of the U.S. farmland. Some specific states where zone tillage is currently in practice are Pennsylvania, Connecticut, Minnesota, Indiana, Wisconsin, and Illinois.

Unfortunately, its use in the Northern Cornbelt states lacks consistent yield results; however, there is still interest in deep tillage within the agriculture industry. In areas that are not well-drained, deep tillage may be used as an alternative to installing more expensive tile drainage.

Effects of Tillage

Rice tillage

Positive

Plowing:

- Loosens and aerates the top layer of soil or horizon A, which facilitates planting the crop.

- Helps mix harvest residue, organic matter (humus), and nutrients evenly into the soil.

- Mechanically destroys weeds.

- Dries the soil before seeding (in wetter climates tillage aids in keeping the soil drier).

- When done in autumn, helps exposed soil crumble over winter through frosting and defrosting, which helps prepare a smooth surface for spring planting.

Negative

- Dries the soil before seeding.

- Soil loses nutrients, like nitrogen and fertilizer, and its ability to store water.

- Decreases the water infiltration rate of soil. (Results in more runoff and erosion since the soil absorbs water more slowly than before).

- Tilling the soil results in dislodging the cohesiveness of the soil particles thereby inducing erosion.

- Chemical runoff.

- Reduces organic matter in the soil.

- Reduces microbes, earthworms, ants, etc.

- Destroys soil aggregates.

- Compaction of the soil, also known as a tillage pan.

- Eutrophication (nutrient runoff into a body of water).

- Can attract slugs, cut worms, army worms, and harmful insects to the leftover residues.

- Crop diseases can be harbored in surface residues.

Alternatives to Tilling

Modern agricultural science has greatly reduced the use of tillage. Crops can be grown for several years without any tillage through the use of herbicides to control weeds, crop varieties that tolerate packed soil, and equipment that can plant seeds or fumigate the soil without really digging it up. This practice, called no-till farming, reduces costs and environmental change by reducing soil erosion and diesel fuel usage.

Site Preparation of Forest Land

Site preparation is any of various treatments applied to a site in order to ready it for seeding or planting. The purpose is to facilitate the regeneration of that site by the chosen method. Site preparation may be designed to achieve, singly or in any combination: improved access, by reducing or rearranging slash, and amelioration of adverse forest floor, soil, vegetation, or other biotic factors. Site preparation is undertaken to ameliorate one or more constraints that would otherwise be likely to thwart the objectives of management.

Site preparation is the work that is done before a forest area is regenerated. Some types of site preparation are burning.

Burning

Broadcast burning is commonly used to prepare clearcut sites for planting, e.g., in central British Columbia, and in the temperate region of North America generally.

Prescribed burning is carried out primarily for slash hazard reduction and to improve site conditions for regeneration; all or some of the following benefits may accrue:

- Reduction of logging slash, plant competition, and humus prior to direct seeding, planting, scarifying or in anticipation of natural seeding in partially cut stands or in connection with seed-tree systems.

- Reduction or elimination of unwanted forest cover prior to planting or seeding, or prior to preliminary scarification thereto.

- Reduction of humus on cold, moist sites to favour regeneration.

- Reduction or elimination of slash, grass, or brush fuels from strategic areas around forested land to reduce the chances of damage by wildfire.

Prescribed burning for preparing sites for direct seeding was tried on a few occasions in Ontario, but none of the burns was hot enough to produce a seedbed that was adequate without supplementary mechanical site preparation.

Changes in soil chemical properties associated with burning include significantly increased pH, which Macadam in the Sub-boreal Spruce Zone of central British Columbia found persisting more than a year after the burn. Average fuel consumption was 20 to 24 t/ha and the forest floor depth was reduced by 28% to 36%. The increases correlated well with the amounts of slash (both total and ≥7 cm diameter) consumed. The change in pH depends on the severity of the burn and the amount consumed; the increase can be as much as 2 units, a 100-fold change. Deficiencies of copper and iron in the foliage of white spruce on burned clearcuts in central British Columbia might be attributable to elevated pH levels.

Even a broadcast slash fire in a clearcut does not give a uniform burn over the whole area. Tarrant, for instance, found only 4% of a 140-ha slash burn had burned severely, 47% had burned lightly, and 49% was unburned. Burning after windrowing obviously accentuates the subsequent heterogeneity.

Marked increases in exchangeable calcium also correlated with the amount of slash at least 7 cm in diameter consumed. Phosphorus availability also increased, both in the forest floor and in the 0 cm to 15 cm mineral soil layer, and the increase was still evident, albeit somewhat diminished, 21 months after burning. However, in another study in the same Sub-boreal Spruce Zone found that although it increased immediately after the burn, phosphorus availability had dropped to below pre-burn levels within 9 months.

Nitrogen will be lost from the site by burning, though concentrations in remaining forest floor were found by Macadam to have increased in two out of six plots, the others showing decreases. Nutrient losses may be outweighed, at least in the short term, by improved soil microclimate through the reduced thickness of forest floor where low soil temperatures are a limiting factor.

The *Picea/Abies* forests of the Alberta foothills are often characterized by deep accumulations of organic matter on the soil surface and cold soil temperatures, both of which make reforestation difficult and result in a general deterioration in site productivity; Endean and Johnstone describe experiments to test prescribed burning as a means of seedbed preparation and site amelioration on representative clear-felled *Picea/Abies* areas. Results showed that, in general, prescribed burning did not reduce organic layers satisfactorily, nor did it increase soil temperature, on the sites tested. Increases in seedling establishment, survival, and growth on the burned sites were probably the result of slight reductions in the depth of the organic layer, minor increases in soil temperature, and marked improvements in the efficiency of the planting crews. Results also suggested that the process of site deterioration has not been reversed by the burning treatments applied.

Ameliorative Intervention

Slash weight (the oven-dry weight of the entire crown and that portion of the stem less than four inches in diameter) and size distribution are major factors influencing the forest fire hazard on harvested sites. Forest managers interested in the application of prescribed burning for hazard reduction and silviculture, were shown a method for quantifying the slash load by Kiil. In west-central Alberta, he felled, measured, and weighed 60 white spruce, graphed (a) slash weight per merchantable unit volume against diameter at breast height (dbh), and (b) weight of fine slash (<1.27 cm) also against dbh, and produced a table of slash weight and size distribution on one acre of a hypothetical stand of white spruce. When the diameter distribution of a stand is unknown, an estimate of slash weight and size distribution can be obtained from average stand diameter, number of trees per unit area, and merchantable cubic foot volume. The sample trees in Kiil's study had full symmetrical crowns. Densely growing trees with short and often irregular crowns would probably be overestimated; open-grown trees with long crowns would probably be underestimated.

The need to provide shade for young outplants of Engelmann spruce in the high Rocky Mountains is emphasized by the U.S. Forest Service. Acceptable planting spots are defined as microsites on the north and east sides of down logs, stumps, or slash, and lying in the shadow cast by such material. Where the objectives of management specify more uniform spacing, or higher densities, than obtainable from an existing distribution of shade-providing material, redistribution or importing of such material has been undertaken.

Access

Site preparation on some sites might be done simply to facilitate access by planters, or to improve access and increase the number or distribution of microsites suitable for planting or seeding.

Wang et al. determined field performance of white and black spruces 8 and 9 years after outplanting on boreal mixedwood sites following site preparation (Donaren disc trenching versus no trenching) in 2 plantation types (open versus sheltered) in southeastern Manitoba. Donaren trenching slightly reduced the mortality of black spruce but significantly increased the mortality of white spruce. Significant difference in height was found between open and sheltered plantations for black spruce but not for white spruce, and root collar diameter in sheltered plantations was significantly larger than in open plantations for black spruce but not for white spruce. Black spruce open plantation had significantly smaller volume (97 cm^3) compared with black spruce sheltered (210 cm^3), as well as white spruce open (175 cm^3) and sheltered (229 cm^3) plantations. White spruce open plantations also had smaller volume than white spruce sheltered plantations. For transplant stock, strip plantations had a significantly higher volume (329 cm^3) than open plantations (204 cm^3). Wang et al. recommended that sheltered plantation site preparation should be used.

Mechanical

Up to 1970, no "sophisticated" site preparation equipment had become operational in Ontario, but the need for more efficacious and versatile equipment was increasingly recognized. By this time, improvements were being made to equipment originally developed by field staff, and field testing of equipment from other sources was increasing.

According to J. Hall, in Ontario at least, the most widely used site preparation technique was post-harvest mechanical scarification by equipment front-mounted on a bulldozer (blade, rake, V-plow, or teeth), or dragged behind a tractor (Imsett or S.F.I. scarifier, or rolling chopper). Drag type units designed and constructed by Ontario's Department of Lands and Forests used anchor chain or tractor pads separately or in combination, or were finned steel drums or barrels of various sizes and used in sets alone or combined with tractor pad or anchor chain units.

J. Hall's report on the state of site preparation in Ontario noted that blades and rakes were found to be well suited to post-cut scarification in tolerant hardwood stands for natural regeneration of yellow birch. Plows were most effective for treating dense brush prior to planting, often in conjunction with a planting machine. Scarifying teeth, e.g., Young's teeth, were sometimes used to prepare sites for planting, but their most effective use was found to be preparing sites for seeding, particularly in backlog areas carrying light brush and dense herbaceous growth. Rolling choppers found application in treating heavy brush but could be used only on stone-free soils. Finned drums were commonly used on jack pine–spruce cutovers on fresh brushy sites with a deep duff layer and heavy slash, and they needed to be teamed with a tractor pad unit to secure good distribution of the slash. The S.F.I. scarifier, after strengthening, had been "quite successful" for 2 years, promising trials were under way with the cone scarifier and barrel ring scarifier, and development had begun on a new flail scarifier for use on sites with shallow, rocky soils. Recognition of the need to become more effective and efficient in site preparation led the Ontario Department of Lands and Forests to adopt the policy of seeking and obtaining for field testing new equipment from Scandinavia and elsewhere that seemed to hold promise for Ontario conditions, primarily in the north. Thus, testing was begun of the Brackekultivator from Sweden and the Vako-Visko rotary furrower from Finland.

Mounding

Site preparation treatments that create raised planting spots have commonly improved outplant performance on sites subject to low soil temperature and excess soil moisture. Mounding can certainly have a big influence on soil temperature. Draper et al., for instance, documented this as well as the effect it had on root growth of outplants.

The mounds warmed up quickest, and at soil depths of 0.5 cm and 10 cm averaged 10 and 7 °C higher, respectively, than in the control. On sunny days, daytime surface temperature maxima on the mound and organic mat reached 25 °C to 60 °C, depending on soil wetness and shading. Mounds reached mean soil temperatures of 10 °C at 10 cm depth 5 days after planting, but the control did not reach that temperature until 58 days after planting. During the first growing season, mounds had 3 times as many days with a mean soil temperature greater than 10 °C than did the control microsites.

Draper et al.'s mounds received 5 times the amount of photosynthetically active radiation (PAR) summed over all sampled microsites throughout the first growing season; the control treatment consistently received about 14% of daily background PAR, while mounds received over 70%. By November, fall frosts had reduced shading, eliminating the differential. Quite apart from its effect on temperature, incident radiation is also important photosynthetically. The average control microsite was exposed to levels of light above the compensation point for only 3 hours, i.e., one-quarter of the daily light period, whereas mounds received light above the compensation point for 11 hours, i.e., 86% of the same daily period. Assuming that incident light in the 100-600 $\mu Em^{-2}s^{-1}$ intensity range is the most important for photosynthesis, the mounds received over 4 times the total daily light energy that reached the control microsites.

Orientation of Linear Site Preparation

With linear site preparation, orientation is sometimes dictated by topography or other considerations, but the orientation can often be chosen. It can make a difference. A disk-trenching experiment in the Sub-boreal Spruce Zone in interior British Columbia investigated the effect on growth of young outplants (lodgepole pine) in 13 microsite planting positions: berm, hinge, and trench in each of north, south, east, and west aspects, as well as in untreated locations between the furrows. Tenth-year stem volumes of trees on south-, east-, and west-facing microsites were significantly greater than those of trees on north-facing and untreated microsites. However, planting spot selection was seen to be more important overall than trench orientation.

In a Minnesota study, the N–S strips accumulated more snow but snow melted faster than on E–W strips in the first year after felling. Snow-melt was faster on strips near the centre of the strip-felled area than on border strips adjoining the intact stand. The strips, 50 feet (15.24 m) wide, alternating with uncut strips 16 feet (4.88 m) wide, were felled in a *Pinus resinosa* stand, aged 90 to 100 years.

ORGANIC NO-TILL

Since the advent of no-till in conventional row crop production, soil conservation and improvement aspects of no-tillage systems have attracted the interest of some organic farmers. The big question, of course, is how to do it without synthetic herbicides. The first big breakthrough occurred in the 1980s with the discovery that certain winter annual cover crops, notably cereal rye and hairy vetch, can be killed by mowing at a sufficiently late stage in their development—full head emergence with pollen shed in cereal grains, and full bloom in legumes. When winter annual cover crops at this stage of development are cut close to the ground, they generally do not regrow significantly, and the clippings form an in situ mulch through which vegetables can be transplanted with no or minimal tillage. The mulch hinders weed seed germination and seedling emergence, often for several weeks. This strategy is sometimes called "organic no-till," although continuous no-till is usually not feasible at this time for organic production of annual crops.

Summer squash (left) and broccoli (right) were planted no-till into a winter rye–hairy vetch mulch. The cover crop generated over three tons aboveground dry weight per acre, and was either mechanically rolled (left) or manually cut with a sickle (right) and left in place, providing sufficient mulch to suppress weeds throughout the crops' minimum weed-free periods. No postplant weeding, cultivation, or herbicide applications were done in these crops.

In some initial experiments, the mulch effect of the mechanically killed cover crop was sufficient to delay the onset of weed growth until after the crop's minimum weed-free period, which made postplant cultivation, herbicides, or hand weeding unnecessary. Crop yields were commensurate with yields in control treatments in which the cover crop was incorporated. Tomato and some late-spring brassica plantings did especially well, and some large-seeded crops can be successfully direct-sown into cover crop residues. Such results, combined with research findings that no-tillage systems with cover crops can substantially rebuild soil organic matter and soil quality, stimulated widespread interest in developing organic no-tillage systems.

Challenges in Organic No-till

Several problems dampened the initial enthusiasm for organic no-till. First, results were inconsistent, and both weed control and vegetable yields sometimes fell short of the standard set by vegetables planted after cover crops were tilled in as green manures. Weed suppression failed particularly when cover crop biomass was insufficient to provide a thick mulch, or when many perennial weeds were present.

In addition, planting vegetables through mulch can delay vegetable growth and maturity, or promote problems with slugs, cut worms, or certain crop diseases. Lower soil temperature under the mulch slows mineralization of soil nitrogen (N), which can reduce yields of some crops, especially broccoli and cool season greens, which need a lot of N over a relatively short period of time. Furthermore, the randomly-oriented cover crop residues left by most mowers, scythes, and other manual cutting tools, interfere with mechanical no-till transplanting, and most standard vegetable planters and transplanters do not function well in untilled soil.

Figure: Canada thistle, an invasive perennial weed, easily emerged through a cover crop mulch to compete with this no-till planted broccoli. The farmer has since brought this infestation under control through a combination of timely tillage and high biomass cover crops.

Finally, even when the cover crop residues eliminate early-season weed competition without hindering crop yields, later-season weeds almost invariably emerge, making tillage necessary after vegetable harvest to prepare a seedbed for the following cover crop. Thus continuous organic no-till is generally not practical, and some tillage is usually needed at least once per calendar year to manage weeds in organic annual cropping systems.

Figure: These winter cover crops were cut manually with a sickle prior to manual planting of the broccoli crop shown in figure. Mechanical transplanters, even those designed for no-till applications cannot operate effectively in randomly oriented residues like these.

Advances in Organic No-till

In recent years, technical advances, combined with additional success stories from some working organic farms, have stimulated continued experimentation by farmers and researchers across the United States. No-till transplanters, seeders, and planting aids (coulter followed by a shank to prepare a narrow, deep slot of loosened soil for planting vegetables) have been developed that permit mechanized applications. Whereas the randomly oriented residues of sicklebar- or rotary-mowed cover crops may cause clogging problems with these implements, other options such as flail mowing, rolling, and roll-crimping show promise. The flail mower chops the residue fine enough for the no-till planters to penetrate the mulch to plant seeds or starts into the soil, though the finely chopped residues break down faster so that their weed suppression effect is shorter-lived. Rolled residues are oriented parallel to the direction of travel, thereby minimizing interference with mechanized planting.

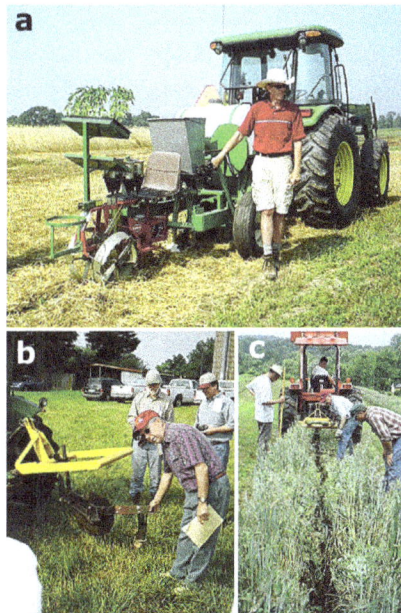

(a) Dr. Keith Baldwin of North Carolina Agriculture and Technology State University demonstrates a no-till vegetable transplanter, developed By Dr. Ron Morse at Virginia Tech and shown here in operation in a cover crop mulch at NCA&TSU's field station in Greensboro, NC. The planter parts residues, loosens soil in a narrow slot to 6–8 inches, plants and waters seedlings, and firms soil around seedlings. It is a major capital investment (at least $7000 per row). (b) Dr. Morse demonstrates a simple, inexpensive, light-duty no-till planting aid consisting of a coulter and shank to part residues and prepare a slot into which seedlings or large seed can be set manually or with standard planting equipment. (c) A no-till planting aid operating through still-standing cover crop prior to planting potatoes or large seeded vegetables. The cover crop is then mown just before the vegetable emerges to suppress weeds for maximum duration during the vegetable's growth.

(a) Cover crop residues that have either been chopped into fairly small pieces by a flail mower, or rolled so that their stems are oriented parallel to the direction of travel, permit mechanized no-till vegetable planting or transplanting. The flail mower shown chopping the cover crop here can also be drawn through the field with the PTO turned off to roll the cover crop (left foreground); thus, it is a versatile tool for no-till cover crop management. (b) Flail mower (PTO off) rolling a dense stand of winter rye–hairy vetch to form a thick mulch with residues oriented to allow mechanized planting.

Researchers have also found that winter cover crops are not the only ones amenable to no-till, no-herbicide management. Oats, field peas, and some vetches planted in early spring can be killed by mowing or rolling in early summer; some warm-season cover crops like soybean, foxtail millet, pearl millet, and sunnhemp can be similarly terminated about two months after seeding. The soil cooling effects of these mulches for midsummer and early fall vegetable plantings can be advantageous during hot seasons.

This sudangrass cover crop, planted in July on Cape Cod, MA, winter-killed to form an in situ mulch that suppressed winter weed growth. This photo was taken in February, at which time common chickweed had begun to cover unmulched fallow beds.

Non-winter-hardy cover crops planted 60–90 days prior to anticipated frost-kill can form an in situ mulch that suppresses winter weeds and lasts into the following spring. While attempts at no-till spring vegetable planting into winterkilled cover crops have been hampered by slow soil warming and/or inadequate weed suppression later in the spring, reduced tillage methods (strip till, ridge till, or shallow tillage) have given more promising results, with good vegetable yields and less problems with weeds compared to the no-till option.

Although continuous organic no-till does not yet appear feasible, significant opportunities exist to reduce tillage in organic production, thereby conserving soil organic matter and soil quality, and possibly improving weed management. Within this context, no-till cover crop management and vegetable planting can be effective tactics when used at certain stages of a crop rotation.

By minimizing soil disturbance and exposure of weed seeds to light, rolling or mowing a cover crop rather than tilling it in helps to close the niche for annual weeds between a cover crop and the subsequent vegetable. Furthermore, surface residues of allelopathic cover crops like rye can maintain a shallow, weed-suppressive zone in the top inch or so of soil for some weeks. Larger seeded crops like beans, peas, and sweet corn, and especially transplanted vegetables, are quite tolerant to the allelopathic effects; hence the cover crop mulch acts to some degree as a selective herbicide. Transplanted tomatoes are especially tolerant to rye allelopathy, and respond positively to substances released by hairy vetch residues. In several studies, tomatoes grown no-till in rye–vetch have out-yielded tomatoes in other systems, including plasticulture with drip irrigation.

For small-seeded vegetable crops that are sensitive to allelopathy and require a fine seedbed, cover crops can be managed with ridge tillage, or strip tillage with sweeps ahead of the tillage implement to clear residues from crop rows. This provides a clean seedbed within crop rows and leaves the mulch between rows to suppress weeds. Ridge or strip tillage also allows prompt soil warming within crop rows, which is important when the farmer plants warm season vegetables early in order to capture early, lucrative markets.

Figure: Strip tillage prepares a narrow (~8 inch) swath through a cover crop mulch, allowing soil warming within the crop row while maintaining weed-suppressive mulch between crop rows.

Considerations in Deciding whether to Attempt Organic No-till

No-till cover crop management and vegetable planting are most likely to work well in fields where:

- The cover crop is mature (heading/flowering with pollen shed), nearly weed-free, and has developed at least three tons (dry weight) aboveground biomass (solid stand three to four feet tall; cannot see the ground when viewed from above; clippings from one square yard weight about one and a half pounds when thoroughly dried).

- The cover crop includes a cereal grain or other grass that forms persistent mulch.

- Nutrient release from mulch decomposition approximately parallels vegetable crop nutrient needs, or sufficient supplemental N and other nutrients are provided.

- Weed populations are light to moderate, with predominantly annual broadleaf weeds.

- The soil is of light to medium texture, well-drained, and quick to warm up in spring.

- Moisture levels are adequate but not excessive, or can be supplemented by in-row drip irrigation.

- Organic mulches have historically been found to harbor natural enemies of important vegetable pests.

- Transplanted or large-seeded vegetables will be planted.

- Early vegetable maturity is not required.

- The farm has access to suitable equipment for rolling/mowing the cover crop and no-till planting through heavy residues.

No-till cover crop management and vegetable planting are not recommended when:

- The cover crop is not mature, is weedy (more than 5–10% of the aboveground biomass is weeds), or has not developed sufficient biomass (looks thin, can see patches of bare ground when viewed from above, dry clippings from one square yard weigh less than one pound).

- The cover crop is likely to decompose too rapidly to provide weed suppression (buckwheat or all-legume cover crops usually break down rapidly).

- Slower N mineralization in the mulched, untilled soil is likely to cause N deficiency in the vegetable crop.

- Invasive perennial weeds are present, the soil's weed seed bank is large, or annual weed populations are high and dominated by grasses or large seeded species that readily penetrate mulch.

- The field has been converted from sod to annual production within the past 12 months (bits of sod can regenerate and become perennial weeds under no-till without herbicides).

- The soil is heavy or clayey, and tends to be slow-draining or slow to warm up, especially in wetter-than-normal years.

- Slugs, squash bugs, or other pests that are commonly associated with organic mulch have historically been a problem.

- Small-seeded vegetables will be direct-sown (sensitive to allelopathy, slugs, and other mulch effects).

- Quick maturation of tomatoes or other vegetables is desired for an early market.

- The farm does not have the equipment needed for planting vegetables through a rolled or mowed cover crop at the scale of operation.

Compared to green manuring (soil incorporation of cover crops), no-till cover crop management slows the rate of N mineralization by keeping the soil cooler, and by leaving the organic residues on the soil surface. This can lead to N deficiency in a cool-season, heavy-feeding crop like spinach or broccoli that requires a lot of N in a short time, especially in heavier soils. On the other hand, the slower N mineralization under a no-till managed cover crop can be advantageous for summer vegetable production on sandy soils in warm climates. In one trial, summer squash yield doubled in mow-killed rye–vetch compared to squash grown after the cover crop was tilled in. The farmer suspects that the no-till system released N roughly as the crop required it, whereas much of the N released from the tilled cover crop leached away before the squash could utilize it.

Terminating cover crops by mowing or rolling should not be described as an organic no-tillage system but as a tool or tactic that can be a valuable component of ecological weed management when used under the right circumstances. It can also be considered part of an organic reduced-tillage system, which is a more realistic goal.

References

- Lindbo, Hayes, Adewunmi (2012). Know Soil Know Life: Physical Properties of Soil and Soil Formation. Soil Science Society of America. P. 17. ISBN 9780891189541

- Soil-the-foundation-of-agriculture, library, knowledge, scitable: nature.com, Retrieved 25 August, 2019

- Baveye, P.; Jacobson, A.R.; Allaire, S.E.; Tandarich, J.P.; Bryant, R.B. (2006). "Whither Goes Soil Science in the United States and Canada?". Soil Science. 171 (7): 501–518. Doi:10.1097/01.ss.0000228032.26905.a9

- Integrated-nutrient-management-inm-meaning-concept-and-goals, soil-fertility: soilmanagementindia.com, Retrieved 31 July, 2019

- F. Stuart Chapin III; Pamela A. Matson; Harold A. Moon (2002). Principles of Terrestrial Ecosystem Ecology. Springer. ISBN 0387954392

- Basic-agronomy-facts, why-fertilizer, crop-nutrition: yara.com, Retrieved 1 August, 2019

- Staff (2007-07-12). "Livestock Antibiotics Can End Up in Human Foods". ENS Newswire. Retrieved 2012-11-14

- "Elephant and cow manure for making paper sustainably" (Press release). Science Daily. March 21, 2018. Retrieved March 30, 2018

5

Agroecology

Agroecology is concerned with the study of ecological processes that are applied to agricultural production systems. Some of the diverse approaches towards agroecology are agro-population ecology, indigenous agroecology and inclusive agroecology. The topics elaborated in this chapter will help in gaining a better perspective about these approaches towards agroecology.

There are three typical ways to define agroecology: As a set of farming practices, as a scientific discipline and as a social movement.

Farming: Agroecological practices are based on ecological inputs and processes, as well as the provision of ecosystem services. Agroecological practices contribute to the different goals of sustainable agriculture: to provide sufficient food for a growing world population, not to be harmful to the environment and natural resources, to limit use of non-renewable energy, and to ensure economic viability for farmers and their communities. Organic farming, diversified crop rotations, biological pest control, extensive agro-pastoral systems and agroforestry are examples of farming method using agroecology.

Science: Increasingly, scientific disciplines and networks express concern about dwindling, finite resources such as fossil fuels, about (related) issues such as climate change, about soil, biodiversity, health and mores.

Although 'business-as-usual' is no longer an option, much mainstream agricultural thinking is focused on retaining the high input industrialised type of farming exemplified by ideas like 'sustainable intensification'.

As a scientific discipline, agroecology studies are quite holistic: they study agroecosystems through an interdisciplinary lens looking at issues such as productivity, stability, sustainability and equitability. They consider issues related to agronomy,ecology, sociology, economics and politics at all relevant scales from the local level to the global level.

Social Movements: Many organisations, as well as many loosely networked individuals are working towards an agro-ecological food and farming future. Taken together, these can be seen, broadly speaking, as a social movement trying to make agri-food more resource savvy and thus genuinely sustainable in the longer term, more people and environment focused. These include the 150+ organisations who have signed up to the ARC2020 platform, large international organisations like Friends of the Earth, Slow Food and IFOAM, national organisations and Individual farmers and consumers: all can and do contribute to an agroecological future.

So instead of the conventional, monoculture-based industrial approach which relies on external inputs, we need to develop sustainable, regenerative farming systems that improve the well-being of small-scale farmers, create diversity to make food production resilient to a changing and unpredictable climate, and produce sufficient food whilst enhancing biodiversity. Instead of marginalising sustainable local food producers, we need to put sustainable local food at the centre of our food supply, with small-scale producers feeding local communities, rather than being squeezed by industrial-scale global supply chains.

Agroecological farming is needed to preserve natural resources. This includes recycling nutrients and energy on the farm, rather than using external inputs; integrating crop and livestock farming; diversifying species (and therefore genetic resources); and focusing on the ways in which crops and livestock can mutually benefit each other, rather than on individual species. By using organic matter and improving the soil, farmers can promote better plant growth.This is an agro-ecology knowledge-intensive system, but the knowledge is developed by the farmer through understanding local conditions and experimenting.

Re-connecting farmers and consumers is important to help building vibrant local food economies. The aim is to support local producers, processors and retailers, and build links between consumers, local farmers and local food businesses. This means creating decentralized short supply chains, diversified markets based on solidarity and fair prices, and closer links between producers and consumers locally. Consumers should be able to purchase ecologically-produced food from small-scale producers. Short distance distribution models are also an important aspect for the closure of nutrient cycles, a basic need in agro-ecological farming practices. To return plant nutrients back into the loop, back to the soil, on the right spot, in the right composition and in the right amounts, is a complex issue. This complexity increases significantly over distance, so agroecology promotes closed production loop sand minimised external inputs. In this way local food economies answer the basic need for plant nutrients in agroecological farming practices.

There are a myriad of different systems offering 'local food' and 'short supply chains' in Europe, including farmers' markets, 'farm-gate' sales, box delivery schemes, mobile shops, community supported agriculture, consumer-producer cooperatives and collective catering and canteens. Short supply chains are not just about reducing the number of intermediaries, about putting the consumer and the producer at the heart of deciding what is produced, how it is produced, and how to define the value.

Food distribution through short supply chains in local markets have been shown to increase income for producers, add value and generate greater autonomy for farmers, and to strengthen local economies by supporting more small businesses. This can improve the viability of small farms, reduces the carbon footprint from food distribution, and enhances household food security by giving people on low income access to good food and healthy diets, as well as encouraging stronger producer-consumer relationships.

Local food supply chains also create employment in rural areas and bring farmers into direct contact with consumers, encouraging the circulation of revenue locally, all the while enhancing social cohesion and making it more likely that farmers can stay farming. This helps foster a sense of community in rural areas, improving quality of life. It can also provide a basis for education on sustainability and ethical issues in urban areas.

Some Relevant Areas for Agroecology

Production Methods

Sustainability and diversity of farming systems, including:

- Re-connect crop and animal production in order to close nutrient cycles.

- Recycle biomass, optimise and close nutrient cycles and reduce dependence external inputs.

- Improve soil conditions, in particular improving organic matter content and biological activity of the soil.

- Integrate protection of biodiversity with production of food and promote and conserve the genetic diversity of crops and animals.

- Minimise resource losses by managing the micro-climate, increasing soil cover, water harvesting.

Processing and Distribution in an Agroecology Framework

- Relocalised and regionalised agroecological food systems that allow fair prices, create jobs and reconnect consumers to farmers.

- foods involving a minimum of industrialised inputs and processes.

- Decentralised and innovative community-led local development that empower people in food production and agroecology (access to land, CSAs, rural development, LEADER).

Participation and Decision-making

- Investigate existing power relations, decision-making processes and opportunities for participation in food systems. Strenghten the role of citizens and consumers in food systems.

- Valorise the diversity of knowledge (local / traditional know-how and practices, common and expert knowledge) in the definition of research problems, the definition of people concerned, and in finding solutions.

- Community-based participatory research and innovation which will facilitate the development of diversified seeds and ecological production and distribution systems (by developing meaningful inter-disciplinary networks, involving a wide range of stakeholders to integrate local and traditional knowledge with formal scientific knowledge).

- Acknowledge the similarities and linkages between agricultural systems in the global North and South. The transition towards sustainable food systems demands integrated and simultaneous solutions in North and South.

APPROACHES TO AGROECOLOGY

Agroecologists do not always agree about what agroecology is or should be in the long-term. Different definitions of the term agroecology can be distinguished largely by the specificity with which one defines the term "ecology", as well as the term's potential political connotations. Definitions of agroecology, therefore, may be first grouped according to the specific contexts within which they situate agriculture. Agroecology is defined by the OECD as "the study of the relation of agricultural crops and environment." This definition refers to the "-ecology" part of "agroecology" narrowly as the natural environment. Following this definition, an agroecologist would study agriculture's various relationships with soil health, water quality, air quality, meso- and micro-fauna, surrounding flora, environmental toxins, and other environmental contexts.

A more common definition of the word can be taken from Dalgaard et al., who refer to agroecology as the study of the interactions between plants, animals, humans and the environment within agricultural systems. Consequently, agroecology is inherently multidisciplinary, including factors from agronomy, ecology, sociology, economics and related disciplines. In this case, the "-ecology" portion of "agroecology is defined broadly to include social, cultural, and economic contexts as well. Francis et al. also expand the definition in the same way, but put more emphasis on the notion of food systems.

Agroecology is also defined differently according to geographic location. In the global south, the term often carries overtly political connotations. Such political definitions of the term usually ascribe to it the goals of social and economic justice; special attention, in this case, is often paid to the traditional farming knowledge of indigenous populations. North American and European uses of the term sometimes avoid the inclusion of such overtly political goals. In these cases, agroecology is seen more strictly as a scientific discipline with less specific social goals.

Agro-population Ecology

This approach is derived from the science of ecology primarily based on population ecology, which over the past three decades has been displacing the ecosystems biology of Odum. Buttel explains the main difference between the two categories, saying that "the application of population ecology to agroecology involves the primacy not only of analyzing agroecosystems from the perspective of the population dynamics of their constituent species, and their relationships to climate and biogeochemistry, but also there is a major emphasis placed on the role of genetics."

Indigenous Agroecology

This concept was proposed by political ecologist Josep Garí to recognise and uphold the integrated agro-ecological practices of many indigenous peoples, who simultaneously and sustainably safeguard, manage and use ecosystems for agricultural, food, biodiversity and cultural purposes at the same time. Indigenous agroecologies are not systems and practices halted in time, but keep co-evolving with new knowledge and resources, such as that provided by development projects, research initiatives and agro-biodiversity exchanges. In fact, the first agro-ecologists were indigenous peoples that advocated development policies and programmes to support their systems, rather than replacing them.

Inclusive Agroecology

Rather than viewing agroecology as a subset of agriculture, Wojtkowski takes a more encompassing perspective. In this, natural ecology and agroecology are the major headings under ecology. Natural ecology is the study of organisms as they interact with and within natural environments. Correspondingly, agroecology is the basis for the land-use sciences. Here humans are the primary governing force for organisms within planned and managed, mostly terrestrial, environments.

As key headings, natural ecology and agroecology provide the theoretical base for their respective sciences. These theoretical bases overlap but differ in a major way. Economics has no role in the functioning of natural ecosystems whereas economics sets direction and purpose in agroecology.

Under agroecology are the three land-use sciences, agriculture, forestry, and agroforestry. Although these use their plant components in different ways, they share the same theoretical core.

Beyond this, the land-use sciences further subdivide. The subheadings include agronomy, organic farming, traditional agriculture, permaculture, and silviculture. Within this system of subdivisions, agroecology is philosophically neutral. The importance lies in providing a theoretical base hitherto lacking in the land-use sciences. This allows progress in biocomplex agroecosystems including the multi-species plantations of forestry and agroforestry.

Ecosystems Agroecology

This approach is driven by the ecosystems biology of Eugene Odum. This approach is based in the hypotheses that the natural systems, with its stability and resilience, provide the best model to mimic if sustainability is the goal. Normally, ecosystems agroecology is not actively involved in social science; however, this school is essentially based on the belief that large-scale agriculture is inappropriate. The work of Steve Gliessman is prototypical of this approach.

Agronomic Ecology

The basic approach in this branch is derived mostly from agronomy, including the traditional agricultural production sciences. This approach also does not actively involve social sciences in the agroecological analysis, but uses social sciences to understand the processes by which agriculture became unsustainable. Chuck Francis, Richard Hardwood, Ricardo Salvador, and Matt Liebman are exemplars of this approach.

Ecological Political Economy

The driving force behind this form of agroecology is a political-economical critique of modern agriculture. The school believes that only radical changes in political economy and the moral economy of research will reduce the negative costs of modern agriculture. The works of Miguel Altieri (ecosystem biologist), John Vandermeer (population ecologist), Richard Lewontin, and Richard Levins provide examples of this politically charged and socially-oriented version of agroecology.

Integrated Assessment of Multifunctional Agricultural Systems

This approach focuses on the multifunctionality of the landscape, instead of focusing solely on the agricultural enterprise. Agriculture and the food system are considered parts of an institutional complex that relates to and integrates with other social institutions. Scholars adopting this highly integrated approach, mostly Europeans, do not consider any one discipline the leader of agroecology.

Holon Agroecology

First introduced in 2007 by the soil scientist William T. Bland and the environmental sociologist Michael M. Bell of the University of Wisconsin–Madison, holon agroecology draws on Koestler's notion of a "holon" which is both part and whole and develops it with ideas of narrative, intentionality, and incompleteness or unfinalizability, within an ever-changing "ecology of contexts". In contrast to systems thinking, holon agroecology stresses seeing the agricultural endeavor as an unfinished accomplishment that is constantly adjusting itself to its many contexts and their conflicts and incommensurabilities. The farm holon represents a kind of "holding together" in order to persist through change, but a holding together that is never fully unified and worked out.

APPLICATIONS OF AGROECOLOGY

To arrive at a point of view about a particular way of farming, an agroecologist would first seek to understand the contexts in which the farm is involved. Each farm may be inserted in a unique combination of factors or contexts. Each farmer may have their own premises about the meanings of an agricultural endeavor, and these meanings might be different from those of agroecologists. Generally, farmers seek a configuration that is viable in multiple contexts, such as family, financial, technical, political, logistical, market, environmental, spiritual. Agroecologists want to understand the behavior of those who seek livelihoods from plant and animal increase, acknowledging the organization and planning that is required to run a farm.

Organic and Non-organic Milk Production

Because organic agriculture proclaims to sustain the health of soils, ecosystems, and people, it has much in common with Agroecology; this does not mean that Agroecology is synonymous with organic agriculture, nor that Agroecology views organic farming as the 'right' way of farming. Also, it is important to point out that there are large differences in organic standards among countries and certifying agencies.

Three of the main areas that agroecologists would look at in farms, would be: the environmental impacts, animal welfare issues, and the social aspects.

Environmental impacts caused by organic and non-organic milk production can vary significantly. For both cases, there are positive and negative environmental consequences.

Compared to conventional milk production, organic milk production tends to have lower eutrophication potential per ton of milk or per hectare of farmland, because it potentially reduces leaching of nitrates (NO_3^-) and phosphates (PO_4^-) due to lower fertilizer application rates. Because organic milk production reduces pesticides utilization, it increases land use per ton of milk due to decreased crop yields per hectare. Mainly due to the lower level of concentrates given to cows in organic herds, organic dairy farms generally produce less milk per cow than conventional dairy farms. Because of the increased use of roughage and the, on-average, lower milk production level per cow, some research has connected organic milk production with increases in the emission of methane.

Animal welfare issues vary among dairy farms and are not necessarily related to the way of producing milk (organically or conventionally).

A key component of animal welfare is freedom to perform their innate (natural) behavior, and this is stated in one of the basic principles of organic agriculture. Also, there are other aspects of animal welfare to be considered – such as freedom from hunger, thirst, discomfort, injury, fear, distress, disease and pain. Because organic standards require loose housing systems, adequate bedding, restrictions on the area of slatted floors, a minimum forage proportion in the ruminant diets, and tend to limit stocking densities both on pasture and in housing for dairy cows, they potentially promote good foot and hoof health. Some studies show lower incidence of placenta retention, milk fever, abomasums displacement and other diseases in organic than in conventional dairy herds. However, the level of infections by parasites in organically managed herds is generally higher than in conventional herds.

Social aspects of dairy enterprises include life quality of farmers, of farm labor, of rural and urban communities, and also includes public health.

Both organic and non-organic farms can have good and bad implications for the life quality of all the different people involved in that food chain. Issues like labor conditions, labor hours and labor rights, for instance, do not depend on the organic/non-organic characteristic of the farm; they can be more related to the socio-economical and cultural situations in which the farm is inserted, instead.

As for the public health or food safety concern, organic foods are intended to be healthy, free of contaminations and free from agents that could cause human diseases. Organic milk is meant to have no chemical residues to consumers, and the restrictions on the use of antibiotics and chemicals in organic food production has the purpose to accomplish this goal. Although dairy cows in both organic and conventional farming practices can be exposed to pathogens, it has been shown that, because antibiotics are not permitted as a preventative measure in organic practices, there are far fewer antibiotic resistant pathogens on organic farms. This dramatically increases the efficacy of antibiotics when/if they are necessary.

In an organic dairy farm, an agroecologist could evaluate the following:

1. The farm minimizes environmental impacts and increase its level of sustainability, for instance by efficiently increasing the productivity of the animals to minimize waste of feed and of land use.

2. Ways to improve the health status of the herd (in the case of organics, by using biological controls, for instance).

3. Way of farming sustain good quality of life for the farmers, their families, rural labor and communities involved.

No-till Farming

No-tillage is one of the components of conservation agriculture practices and is considered more environmental friendly than complete tillage. There is a general consensus that no-till can increase carbon content of topsoils, especially when combined with cover crops, but whether this improves the function of soils as a carbon sink is contested.

No-till can contribute to higher soil organic matter and organic carbon content in soils, though reports of no-effects of no-tillage in organic matter and organic carbon soil contents also exist, depending on environmental and crop conditions. In addition, no-till can indirectly reduce CO_2 emissions by decreasing the use of fossil fuels.

Most crops can benefit from the practice of no-till, but not all crops are suitable for complete no-till agriculture. Crops that do not perform well when competing with other plants that grow in untilled soil in their early stages can be best grown by using other conservation tillage practices, like a combination of strip-till with no-till areas. Also, crops which harvestable portion grows underground can have better results with strip-tillage, mainly in soils which are hard for plant roots to penetrate into deeper layers to access water and nutrients.

The benefits provided by no-tillage to predators may lead to larger predator populations, which is a good way to control pests (biological control), but also can facilitate predation of the crop itself. In corn crops, for instance, predation by caterpillars can be higher in no-till than in conventional tillage fields.

In places with rigorous winter, untilled soil can take longer to warm and dry in spring, which may delay planting to less ideal dates. Another factor to be considered is that organic residue from the prior year's crops lying on the surface of untilled fields can provide a favorable environment to pathogens, helping to increase the risk of transmitting diseases to the future crop. And because no-till farming provides good environment for pathogens, insects and weeds, it can lead farmers to a more intensive use of chemicals for pest control. Other disadvantages of no-till include underground rot, low soil temperatures and high moisture.

Based on the balance of these factors, and because each farm has different problems, agroecologists will not attest that only no-till or complete tillage is the right way of farming. Yet, these are not the only possible choices regarding soil preparation, since there are intermediate practices such as strip-till, mulch-till and ridge-till, all of them – just as no-till – categorized as conservation tillage. Agroecologists, then, will evaluate the need of different practices for the contexts in which each farm is inserted.

AGROECOSYSTEM

Typical example of artificial ecosystem is a cultivated field or agro-ecosystem. This is a natural system altered by men through agricultural activity.

It's different from a natural ecosystem for four main characteristics:

- Simplification: a farmer favours a plant species removing all other animal or plant species which could damage it.

- The energy intake employed by men in the form of machinery, fertilizers, pesticides, selected seeds, processings.

- The biomass (harvest) which is removed when ripe. This makes the ecosystem an open system, which means it depends from external processes to reintroduce fertilizing substances suitable to nourish a new growth and development process of organic material (plants). A natural ecosystem, instead, self-fertilizes as the biomass remains in its original setting.

- The introduction of pollutant substances which, in the case of intensive agriculture, are chemical fertilizers, antiparasitics and other chemical non biodegradable substances which accumulate in the ecosystem or which seep in the subsoil, in some cases getting to the point of seriously polluting groundwaters, seas and rivers.

A home is also a small artifical ecosystem. Objects, food, solar energy, water, etc. are introduced inside houses from outdoor and solid and liquid waste generated by human activities is removed outdoor. The city functions in the same way. A city, in fact, depends from external areas for water and food supplies as well as building materials and other resources necessary for its development and waste generated in a city is unloaded outside the urban area (in landfills and incinerators), which means everything which doesn't contribute to the survival of the urban ecosystem is deposited in these areas.

AGROFORESTRY

Agroforestry is a collective name for land-use systems and technologies where woody perennials (trees, shrubs, palms, bamboos, etc.) are deliberately used on the same land-management units as agricultural crops and/or animals, in some form of spatial arrangement or temporal sequence. In agroforestry systems there are both ecological and economical interactions between the different components. Agroforestry can also be defined as a dynamic, ecologically based, natural resource management system that, through the integration of trees on farms and in the agricultural landscape, diversifies and sustains production for increased social, economic and environmental benefits for land users at all levels. In particular, agroforestry is crucial to smallholder farmers and other rural people because it can enhance their food supply, income and health. Agroforestry systems are multifunctional systems that can provide a wide range of economic, sociocultural, and environmental benefits.

There are three main types of agroforestry systems:

- Agrisilvicultural systems are a combination of crops and trees, such as alley cropping or homegardens.

- Silvopastoral systems combine forestry and grazing of domesticated animals on pastures, rangelands or on-farm.

- The three elements, namely trees, animals and crops, can be integrated in what are called agrosylvopastoral systems and are illustrated by homegardens involving animals as well as scattered trees on croplands used for grazing after harvests.

Because agroforestry integrates multiple natural components and is at the crossroads of tradition and modernity, it necessarily brings together people from diverse fields of knowledge: agronomists, animal care specialists, landscape planners, foresters, economists, soil analysts and many more. This diversity of disciplines is certainly a strength, but its complexity also represents a challenge, notably in terms of coordination and communication.

Many different words are used to express realities that connect to each other. Terms like climate-smart agriculture and agroecology both incorporate a wide array of practices, and among them is agroforestry. Some practices, such as permaculture, have found a voice in grassroots organizations. In other instances, the emphasis is on integrating trees in agricultural systems, as is the case for evergreen agriculture. These systems all represent a commitment to bringing sustainable development principles to agricultural production. As trees are a fundamental component of many ecosystems, their integration in various farming practices doesn't come as a surprise.

References

- Briefing-note-agroecology, agroecology: arc2020.eu, Retrieved 1 August, 2019

- Funes, Fernando; García, Luis; Bourque, Martin; Pérez, Nilda; Rosset, Peter (January 2002). Sustainable Agriculture and Resistance: Transforming Food Production in Cuba. Oakland, CA: Food First Books. ISBN 978-0-935028-87-4

- Cederlöf, Gustav (2016). "Low-carbon food supply: The ecological geography of Cuban urban agriculture and agroecological theory". Agriculture and Human Values. 33 (4): 771–784. Doi:10.1007/s10460-015-9659-y

- What-is-agroecology, what-is-ecology: ecology.com, Retrieved 1 August, 2019

- Garí, Josep A. (2004). Plant diversity, sustainable rural livelihoods and the HIV/AIDS crisis. Bangkok: UNDP & FAO, 2004. Published in English and Chinese. ISBN 974-92021-4-7

- The-agricultural-ecosystem, what-is-an-ecosystem, argomento: eniscuola.net, Retrieved 28 February, 2019

6

Irrigation and its Types

Irrigation refers to the method of applying controlled amounts of water to plants at needed intervals to grow agricultural crops and maintain landscapes. The diverse methods of irrigation include micro-irrigation, sprinkler irrigation and deficit irrigation. The topics elaborated in this chapter will help in gaining a better perspective about these methods of irrigation.

Irrigation is an artificial application of water to the soil, usually to assist with the growth of crops. In crop production, it is mainly used in dry areas and in periods of rainfall shortfalls, but also to protect plants against frost. Additionally, irrigation helps suppress weed growing in rice fields. By contrast, agriculture that relies only on direct rainfall is referred to as rain-fed farming. Irrigation is often studied together with drainage, which is the natural or artificial removal of surface and sub-surface water from a given area.

Irrigation is also a term used in the medical/dental fields and refers to flushing and washing out anything with water or another liquid.

Manual Irrigation using Buckets or Watering Cans

These systems have low requirements for infrastructure and technical equipment but need high labor inputs. Irrigation using watering cans is to be found for example in peri-urban agriculture around large cities in some African countries.

Automatic, Non-electric Irrigation using Buckets and Ropes

Besides the common manual watering by bucket, an automated, natural version of this also exist. Using plain polyester ropes combined with a prepared ground mixture can be used to water plants from a vessel filled with water. The ground mixture would need to be made depending on the plant itself, yet would mostly consist of black potting soil, vermiculite and perlite. This system would (with certain crops) allow you to save expenses as it does not consume any electricity and only little water (unlike sprinklers, water timers). However, it may only be used with certain crops (probably mostly larger crops that do not need a humid environment).

Irrigation using Stones to Catch Water from Humid Air

In countries where at night, humid air sweeps the countryside, stones are used to catch water from the humid air by condensation. This is for example practiced in the vineyards at Lanzarote.

Dry Terraces for Irrigation and Water Distribution

In subtropical countries as Mali and Senegal, a special type of terracing (without flood irrigation or intent to flatten farming ground) is used. Here, a 'stairs' is made through the use of ground level differences which helps to decrease water evaporation and also distributes the water to all patches (sort of irrigation).

Sources of Irrigation Water

Sources of irrigation water can be groundwater extracted from springs or by using wells, surface water withdrawn from rivers, lakes or reservoirs or non-conventional sources like treated waste-water, desalinated water or drainage water. A special form of irrigation using surface water is spate irrigation, also called floodwater harvesting. In case of a flood (spate) water is diverted to normally dry river beds (wadi's) using a network of dams, gates and channels and spread over large areas. The moisture stored in the soil will be used thereafter to grow crops. Spate irrigation areas are in particular located in semi-arid or arid, mountainous regions. While floodwater harvesting belongs to the accepted irrigation methods, rainwater harvesting is usually not considered as a form of irrigation. Rainwater harvesting is the collection of runoff water from roofs or unused land and the concentration of this water on cultivated land. Therefore this method is considered as a water concentration method.

In-ground Irrigation System

Most commercial and residential irrigation systems are "in ground" systems, which means that everything is buried in the ground. With the pipes, sprinklers, and irrigation valves being hidden, it makes for a cleaner, more presentable landscape without garden hoses or other items having to be moved around manually.

Water Source and Piping

The beginning of a sprinkler system is the water source. This is usually a tap into an existing (city) water line or a pump that pulls water out of a well or a pond. The water travels through pipes from the water source through the valves to the sprinklers. The pipes from the water source up to the irrigation valves are called "mainlines," and the lines from the valves to the sprinklers are called "lateral lines." Most piping used in irrigation systems today are HDPE and MDPE or PVC or PEX plastic pressure pipes due to their ease of installation and resistance to corrosion. After the water source, the water usually travels through a check valve. This prevents water in the irrigation lines from being pulled back into and contaminating the clean water supply.

Controllers, Zones and Valves

Most irrigation systems are divided into zones. A zone is a single Irrigation Valve and one or a group of sprinklers that are connected by pipes. Irrigation Systems are divided into zones because there is usually not enough pressure and available flow to run sprinklers for an entire yard or sports field at once. Each zone has a solenoid valve on it that is controlled via wire by an Irrigation Controller. The Irrigation Controller is either a mechanical or electrical device that signals a zone to turn on at a specific time and keeps it on for a specified amount of time. "Smart Controller" is a recent term used to describe a controller that is capable of adjusting the watering time by itself in response to current environmental conditions. The smart controller determines current conditions by means of historic weather data for the local area, a moisture sensor (water potential or water content), weather station, or a combination of these.

Sprinklers

When a zone comes on, the water flows through the lateral lines and ultimately ends up at the irrigation Sprinkler heads. Most sprinklers have pipe thread inlets on the bottom of them which allows a fitting and the pipe to be attached to them. The sprinklers are usually installed with the top of the head flush with the ground surface. When the water is pressurized, the head will pop up out of the ground and water the desired area until the valve closes and shuts off that zone. Once there is no more water pressure in the lateral line, the sprinkler head will retract back into the ground.

Problems related to Irrigation

- Depletion of underground aquifers. By the middle of the twentieth century, the advent of diesel and electric motors led for the first time to systems that could pump groundwater out of major aquifers faster than it was recharged. This can lead to permanent loss of aquifer capacity, decreased water quality, ground subsidence, and other problems. The future of food production in such areas as the North China Plain, the Punjab, and the Great Plains of the US is threatened.

- Ground subsidence (as in New Orleans, Louisiana).

- Underirrigation gives poor salinity control which leads to increased soil salinity with consequent build up of toxic salts on soil surface in areas with high evaporation. This requires either leaching to remove these salts and a method of drainage to carry the salts away or use of mulch to minimize evaporation.

- Overirrigation because of poor distribution uniformity or management wastes water and chemicals, and may lead to water pollution.

- Deep drainage (from over-irrigation) may result in rising water tables which in some instances will lead to problems of irrigation salinity.

- Irrigation with saline or high-sodium water may damage soil structure.

- Competition for surface water rights.

SURFACE IRRIGATION

Surface irrigation is where water is applied and distributed over the soil surface by gravity. It is by far the most common form of irrigation throughout the world and has been practiced in many areas virtually unchanged for thousands of years.

Surface irrigation is often referred to as flood irrigation, implying that the water distribution is uncontrolled and therefore, inherently inefficient. In reality, some of the irrigation practices grouped under this name involve a significant degree of management (for example surge irrigation). Surface irrigation comes in three major types; level basin, furrow and border strip.

Process

The process of surface irrigation can be described using four phases. As water is applied to the top end of the field it will flow or advance over the field length. The advance phase refers to that length of time as water is applied to the top end of the field and flows or advances over the field length. After the water reaches the end of the field it will either run-off or start to pond. The period of time between the end of the advance phase and the shut-off of the inflow is termed the wetting, ponding or storage phase. As the inflow ceases the water will continue to runoff and infiltrate until the entire field is drained. The depletion phase is that short period of time after cut-off when the length of the field is still submerged. The recession phase describes the time period while the water front is retreating towards the downstream end of the field. The depth of water applied to any point in the field is a function of the opportunity time, the length of time for which water is present on the soil surface.

Types of Surface Irrigation

Basin Irrigation

Level basin flood irrigation on wheat.

Residential flood irrigation in the Southwest, United States of America.

Level basin irrigation has historically been used in small areas having level surfaces that are surrounded by earth banks. The water is applied rapidly to the entire basin and is allowed to infiltrate.

In traditional basins no water is permitted to drain from the field once it is irrigated. Basin irrigation is favoured in soils with relatively low infiltration rates. This is also a method of surface irrigation. Fields are typically set up to follow the natural contours of the land but the introduction of laser levelling and land grading has permitted the construction of large rectangular basins that are more appropriate for mechanised broadacre cropping.

Drainback Level Basins

Drain back level basins (DBLB) or contour basins are a variant of basin irrigation where the field is divided into a number of terraced rectangular bays which are graded level or have no significant slope. Water is applied to the first bay (usually the highest in elevation) and when the desired depth is applied water is permitted to drain back off that bay and flow to the next bay which is at a lower elevation than the first. Each bay is irrigated in turn using a combination of drainage water from the previous bay and continuing inflow from the supply channel. Successful operation of these systems is reliant on a sufficient elevation drop between successive bays. These systems are commonly used in Australia where rice and wheat are grown in rotation.

Furrow Irrigation

Furrow irrigation system using siphon tubes.   Gated pipe supply system.

Furrow irrigation is conducted by creating small parallel channels along the field length in the direction of predominant slope. Water is applied to the top end of each furrow and flows down the field under the influence of gravity. Water may be supplied using gated pipe, siphon and head ditch, or bankless systems. The speed of water movement is determined by many factors such as slope, surface roughness and furrow shape but most importantly by the inflow rate and soil infiltration rate. The spacing between adjacent furrows is governed by the crop species, common spacings typically range from 0.75 to 2 metres. The crop is planted on the ridge between furrows which may contain a single row of plants or several rows in the case of a bed type system. Furrows may range anywhere from less than 100 m to 2000 m long depending on the soil type, location and crop type. Shorter furrows are commonly associated with higher uniformity of application but result in increasing potential for runoff losses. Furrow irrigation is particularly suited to broadacre row crops such as cotton, maize and sugar cane. It is also practiced in various horticultural industries such as citrus, stone fruit and tomatoes.

The water can take a considerable period of time to reach the other end, meaning water has been infiltrating for a longer period of time at the top end of the field. This results in poor uniformity with high application at the top end with lower application at the bottom end. In most cases the performance of furrow irrigation can be improved through increasing the speed at which water moves along the field (the advance rate). This can be achieved through increasing flow rates or through the practice of surge irrigation. Increasing the advance rate not only improves the uniformity but also reduces the total volume of water required to complete the irrigation.

Surge Irrigation

Surge Irrigation is a variant of furrow irrigation where the water supply is pulsed on and off in planned time periods (e.g. on for 1 hour off for 1½ hour). The wetting and drying cycles reduce infiltration rates resulting in faster advance rates and higher uniformity than continuous flow. The reduction in infiltration is a result of surface consolidation, filling of cracks and micro pores and the disintegration of soil particles during rapid wetting and consequent surface sealing during each drying phase. On those soils where surging is effective it has been reported to allow completion of the irrigation with a lower overall water usage and therefore higher efficiency and potentially offer the ability to practice deficit irrigation. The effectiveness of surge irrigation is soil type dependent; for example, many clay soils experience a rapid sealing behaviour under continuous flow and therefore surge irrigation offers little benefit.

Bay/border Strip Irrigation

Border strip, otherwise known as border check or bay irrigation could be considered as a hybrid of level basin and furrow irrigation. The field is divided into a number of bays or strips, each bay is separated by raised earth check banks (borders). The bays are typically longer and narrower compared to basin irrigation and are orientated to align lengthwise with the slope of the field. Typical bay dimensions are between 10-70m wide and 100-700m long. The water is applied to the top end of the bay, which is usually constructed to facilitate free-flowing conditions at the downstream end. One common use of this technique includes the irrigation of pasture for dairy production.

Issues Associated with Surface Irrigation

While surface irrigation can be practiced effectively using the correct management under the right conditions, it is often associated with a number of issues undermining productivity and environmental sustainability:

- Waterlogging - Can cause the plant to shut down delaying further growth until sufficient water drains from the rootzone. Waterlogging may be counteracted by drainage, tile drainage or watertable control by another form of subsurface drainage.

- Deep drainage - Overirrigation may cause water to move below the root zone resulting in rising water tables. In regions with naturally occurring saline soil layers (for example salinity in south eastern Australia) or saline aqifers, these rising water tables may bring salt up into the root zone leading to problems of irrigation salinity.

- Salinization - Depending on water quality irrigation water may add significant volumes of salt to the soil profile. While this is a lesser issue for surface irrigation compared to other irrigation methods (due to the comparatively high leaching fraction), lack of subsurface drainage may restrict the leaching of salts from the soil. This can be remedied by drainage and soil salinity control through flushing.

The aim of modern surface irrigation management is to minimize the risk of these potential adverse impacts.

MICRO IRRIGATION

Micro-irrigation, also called localised irrigation, low volume irrigation, low-flow irrigation, or trickle irrigation is an irrigation method with lower pressure and flow than a traditional sprinkler system. Low volume irrigation is used in agriculture for row crops, orchards, and vineyards. It is also used in horticulture in wholesale nurseries, in landscaping for civic, commercial, and private landscapes and gardens, and in the science and practice of restoration ecology and environmental remediation.

There are several types of micro-irrigation systems. Many of the components are the same for all of these types of systems. Most systems typically include filters, pipes, valves, and tubing. The main difference is in the type of emission device that is used to deliver the water to the plants. Drip irrigation utilizes drip emitters that deliver water at very low rates. The typical range is 0.2 to 4.0 gallons per hour. In some systems, the emitters are installed manually on the outside of the tubing and placed where needed. Other systems might use integral dripperline or drip tape with the emitters already installed at a predetermined spacing. Micro-sprinklers, which can include fixed stream sprays and rotating spinners typically deliver water at a higher rate, such as 10 to 25 gallons per hour and will cover a larger area than drip emitters. These are more typically used in tree orchards where the plants are larger. The goal is to distribute water slowly in small volumes and target it to plants' root zones with less runoff or overspray than landscape and garden conventional spray and rotary sprinklers. The low volume allows the water to penetrate and be absorbed into slow-percolation soils, such as clay, minimizing water runoff.

System Components

There are a wide variety of system components included in a micro-irrigation systems. Most systems include a filter. These may include pre-filters, sand separators, media filters, screen filters, and disc filters. The level of filtration required depends on the size of the emission device and the quality of the water source. A pressure regulator or regulating valve may be required to reduce the system pressure to the desired level. Automatic or manually operated valves will be required to switch from one irrigated section to another. An irrigation controller will be used with automatic systems and may also be needed for backflushing the filter or sand separator. Since water conservation is a frequent reason for choosing micro-irrigation systems, soil moisture sensors, rain shutoff sensors, and sometimes even weather stations may be installed.

Emission Devices

Microtubing

Microtubing is one of the oldest types of drip irrigation devices and was used in greenhouses in the

1970s. It consists of a very small diameter tubing. Flow is regulated purely by the length and diameter of the tubing. Weights or stakes are sometimes attached to the end of the tubing to keep it in place.

Fixed Flow Drip Emitters

Low-flow irrigation systems in gardens using drip apply water through two methods:

- Pre installed small holes in small diameter tubes placed on or below the surface.

- Self Cleaning emitters, in different precipitation rates, pre installed or contractor installed for different rate emitters on same supply line (i.e. trees-higher, perennials-lower). The Flexible supply pipe can be buried either underground or pinned on the surface and buried under.

Low volume irrigation systems often use the two delivery components of drip systems to apply water through small holes in small diameter tubes placed on or below the surface of the field. This is done instead of agricultural surface irrigation and furrow irrigation for vegetables, fruits and berries, and other high-value crops.

Adjustable Drip Emitters

Trickle emitters, also called 'spider sprays,' come in fixed or adjustable radius shapes and diameters, and are installed directly on the flexible supply pipe or on tubing connected to it, and mounted on small stakes. Trickle emitter-'Spider sprays' work well for plants with more fibrous root systems, tree and large shrub basins, and in pots and container gardens - allowing automated watering of plants on decks and patios. Mist emitters can also be used in pot, both on the ground and hanging, with humidity-fog watering for epiphytes and ferns replicating habitats.

In the Horticulture industry, wholesale growers and plant nurseries often use the trickle emitters for 5-US-gallon (19 L) and larger container stock, to automate watering. Attached to longer supply tubing on short stakes, they are easily movable to new containers when stock is moved or sold. Mist emitters are used for propagation, epiphytes, and other plants needing higher humidity.

Micro-sprinklers

Low volume micro-sprinklers may be attached to hard plastic risers or attached to standard sprinkler heads, but are more typically mounted on stakes and attached to small diameter micro-tubing connected to polyethylene tubing with a barbed connector. Some micro-sprinklers have a fixed spray or stream pattern, while others rotate. These are installed above ground and are often used for fruit and nut orchards and vineyards. These systems are expensive, even for large-scale agricultural use, and are predominantly used for high-value crops.

Macro-drip Irrigation

High-volume, low-pressure irrigation systems for container gardening are known as Macro-Drip. A pressure regulator lowers the water pressure to under 30 pounds per square inch while a relatively large diameter hose or pipe delivers the water directly to a sprinkler head. This allows a larger volume of water to reach the flowerpot in a short amount of time, which will then be absorbed into the roots of the plant.

Ecological Restoration and Phytoremediation Projects

Low-flow irrigation systems are used on some native plant habitat restoration and environmental remediation projects. The lower operating pressure can be the only choice for remote locations with wells or small storage tank water sources. It is used in temporary installations during initial establishment periods, and being on the soil surface easily removable with minimal damage to the recovering plant community. An example is its use in riparian zone restoration, and environmental remediation projects using Phytoremediation and Bioremediation techniques.

Water Conservation and Regulations

As municipal and agricultural water supplies become more limited; through increased population demands, droughts, and climate change; city, water district, and state-province level regulations and codes are beginning to encourage, offer rebates with use, or mandate significantly reduced water allowances, at higher costs, that are bringing many water conservation products and techniques both to the forefront and more competitively matched to traditional irrigation system costs.

Use of micro-irrigation systems on green building candidate projects can help them to accumulate points for LEED - (Leadership in Energy and Environmental Design) certification rating and awards.

Subsurface Textile Irrigation

Subsurface Textile Irrigation (SSTI) is a technology designed specifically for subsurface irrigation in all soil textures from desert sands to heavy clays. Use of SSTI will significantly reduce the usage of water, fertilizer and herbicide. It will lower on-going operational costs and, if maintained properly, will last for decades. By delivering water and nutrients directly to the root zone, plants are healthier and have a far greater yield.

It is the only irrigation system that can safely use recycled water or treated water without expensive "polishing" treatment because water never reaches the surface.

A typical subsurface textile irrigation system has an impermeable base layer (usually polyethylene or polypropylene), a drip line running along that base, a layer of geotextile on top of the drip line and, finally, a narrow impermeable layer on top of the geotextile. Unlike standard drip irrigation, the spacing of emitters in the drip pipe is not critical as the geotextile moves the water along the fabric up to 2m from the dripper.

SSTI is installed 15–20 cm below the surface for residential/commercial applications and 30–50 cm for agricultural applications.

How SSTI Works

The systems rely on specific geotextiles to absorb the water from the drippers and to rapidly transport that water via mass flow and capillary action along the geotextile effectively turning those single drippers into billions of emitters. In essence, this enables intimate control of the speed of water delivery so that the capillary action of any soil can be matched (something that is virtually impossible for any other irrigation method including bare drip pipe below the surface). If the capillary action of the soil can be matched with the water delivery, only the minimal amount of water is needed to service the needs of the plants.

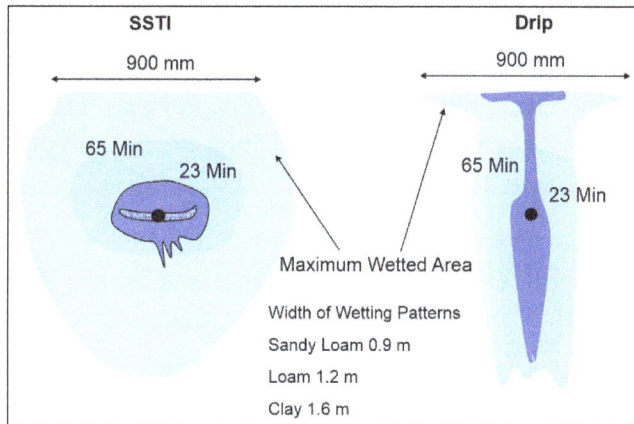

A cross-sectional view of the wetting pattern provided by SSTI, as compared to drip irrigation.

To increase effectiveness, SSTI products should have an impermeable base layer to slow gravitational loss of water and to create an elliptical wetting pattern under the soil surface. It should also have a small impermeable top layer to ensure that water from the dripper does not "tunnel" through the geotextile and up to the surface (again a common problem with bare subsurface drip pipe). The effect of these two layers is dramatic as it maximizes the spread of water through the geotextile (up to 10,000 times faster than a clay loam soil as tested by Charles Sturt University).

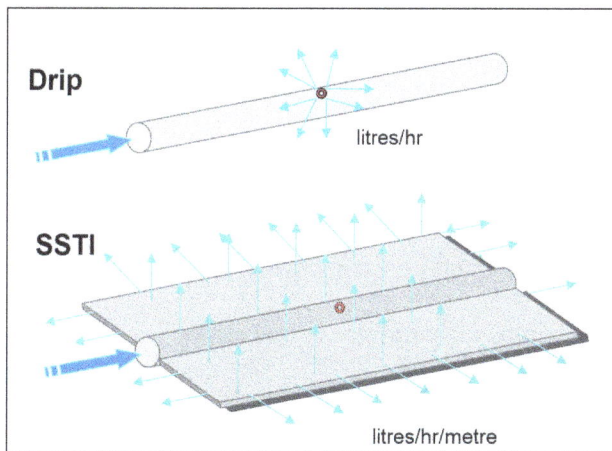

Shows the flow of water through an SSTI installation
as compared to a drip irrigation system.

When comparing SSTI with surface drip, using the same amount of water, SSTI can cover 2.5 times the volume of soil and takes six times longer to dry down until the next irrigation is required.

Recycled Water and Treated Effluent

Recycled water can be used in SSTI systems as it will spread the nutrient load over 2-3 times the soil volume (compared to other irrigation methods). This means that additional nutrient requirements are minimized and the soil will have a long life without overloading other nutrients (especially phosphorus and potassium).

A major benefit of SSTI is that treated effluent can be used but is prevented from reaching the surface. Recreational or agricultural activities can continue on the field during irrigation without the contaminants coming in contact with the public.

Nutrients can be injected through all SSTI systems (fertigation). Macro and micro nutrients can be delivered to specific crops including grass, pasture, trees and vines. The nutrient is placed directly in the root zone so there is almost no wastage and no potential for run off into waterways.

Components

Well-designed SSTI systems are laid out essentially the same as drip systems but in many cases SSTI can be laid in a "serpentine" pattern vastly reducing the number of take-off connections and potential leaks.

Most drip tubes/tapes used in SSTI are pressure compensating. For example, using a 16mm drip tape, a run of up to 180 m can be achieved from one connection. Longer runs of up to 1,000 m can be achieved using lower flow rates per lineal metre and/or larger diameter drip tape.

The following components make up a typical SSTI installation:

- Pump or pressurized water source to 100-300kPa (14-43psi),
- Water filters or filtration system from 120 micron with suspended solids less than 30ppm,
- Fertigation injector systems,
- Back flow prevention,
- Pressure regulating valves,
- Main line can be LDPE or PVC,
- Solenoid valves or gate valves to control water flow,
- SSTI system laterals,
- Barbed or Spinlock fittings with stainless steel clips,
- Flushing valves at the end of laterals or combined laterals into a flushing line so that regular flushing can remove suspended solids or bacteria that may build up when using recycled water.

Using drip tapes in SSTI means that there is a wide range of commercial fittings making SSTI very easy to install. Fittings are usually spinlock or ringlock devices securing the drip tape using a barb.

Performance of SSTI

The advantages of SSTI are:

- SSTI is a "permanent" solution if maintained properly. The components are inert and, given that they are situated below the ground, are not subject to the effects of weather, animals, machinery, vandals or other terrestrial conditions.

- Water savings of 50-75% compared with overhead systems.

- Low pressure requirement (also means lower power requirements).

- Yields can be improved up to four (4) times in certain crops.

- Minimal root intrusion to drippers in the SSTI with a deflective tape on top.

- No emitter blockage due to crusting.

- Minimal effect of evaporation.

- Safe use of recycled or treated water.

- Can use the field (for recreation or agriculture) while irrigation is running.

- Weed growth is minimized because the water does not reach the surface (saving herbicide cost). Germination of weeds only occurs during rainfall.

- Fertigation can be done directly to the root zone (saving fertilizer cost).

- Efficient distribution of nutrients to the entire root zone.

- Broad wetting patterns (moisture covers the entire root zone).

- Water delivery can match the natural capillary rates in soil so saturation is minimized.

- Soil moisture can maintained at field capacity (minimised gravitational losses).

- No surface run off.

- Soil erosion is minimized.

- Fields do not have to be perfectly level.

- Fields with irregular shapes can be accommodated.

- Distance between lines of SSTI is far greater than drip (lower number of solenoids and other components compared to sprinklers and drip).

- SSTI laterals can be ploughed in behind a tractor on large sites (over 10,000 m per day).

- Foliage remains dry (fungal and bacterial leaf disease is minimized).

The disadvantages of SSTI are:

- Initial capital cost is typically more than overhead irrigation.

- Quality installation is critical. If mistakes made, they are difficult to find.

- Installation using correct fittings must be done.

- Regular maintenance is required to ensure long life.

- Automated control and monitoring systems are preferable (subsurface irrigation does not give any visual indicators to show if it is working or not).

- SSTI is usually not UV-treated so it must be kept out of sunlight until it is installed under the surface.

- Temporary overhead watering may be required to establish turf in hot areas.

- Germination of some agricultural crops may require overhead watering if insufficient rainfall.

- SSTI cannot apply fertilizers or herbicides overhead on the surface.

- Rodents may damage the system (although less than drip systems).

Drip Irrigation

Drip irrigation is a type of micro-irrigation system that has the potential to save water and nutrients by allowing water to drip slowly to the roots of plants, either from above the soil surface or buried below the surface. The goal is to place water directly into the root zone and minimize evaporation. Drip irrigation systems distribute water through a network of valves, pipes, tubing, and emitters. Depending on how well designed, installed, maintained, and operated it is, a drip irrigation system can be more efficient than other types of irrigation systems, such as surface irrigation or sprinkler irrigation.

Significance

Modern drip irrigation has arguably become the world's most valued innovation in agriculture since the invention in the 1930s of the impact sprinkler, which offered the first practical alternative to surface irrigation.

Current Developments

Careful study of all the relevant factors like land topography, soil, water, crop and agro-climatic conditions are needed to determine the most suitable drip irrigation system and components to be used in a specific installation.

Micro-spray Heads

Drip irrigation may also use devices called micro-spray heads, which spray water in a small area, instead of dripping emitters. These are generally used on tree and vine crops with wider root zones.

Subsurface Drip Irrigation

Subsurface drip irrigation (SDI) uses permanently or temporarily buried dripperline or drip tape located at or below the plant roots. It is becoming popular for row crop irrigation, especially in areas where water supplies are limited, or recycled water is used for irrigation.

Global Reach and Market Leaders

As of 2012, China and India were the fastest expanding countries in the field of drip- or other micro-irrigation, while worldwide well over ten million hectares utilised these technologies. Still, this amounted to less than 4 percent of the world's irrigated land. That year, Israel's Netafim was the global market leader (a position it maintained in 2018), with India's Jain Irrigation being the second-biggest micro-irrigation company.

Components and Operation

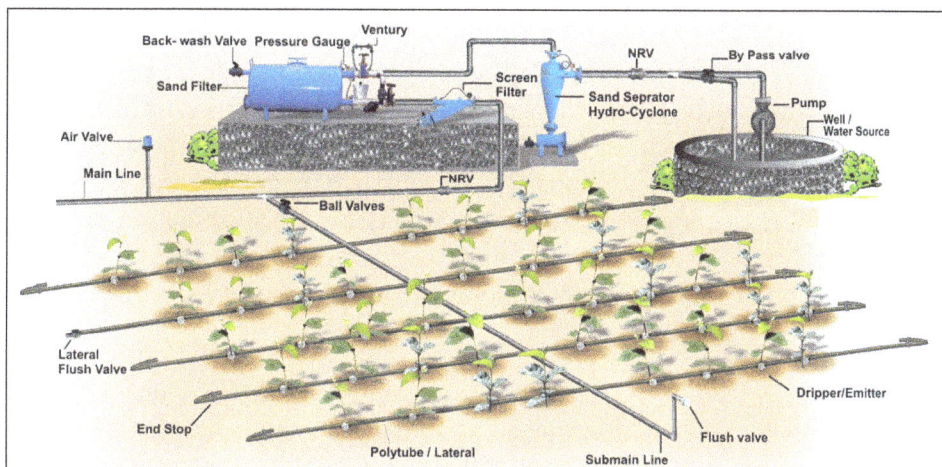
Drip irrigation system layout and its parts.

Nursery flowers watered with drip irrigation in Israel.

Horticulture drip emitter in a pot.

Components used in drip irrigation include:

- Pump or pressurized water source.

- Water filter(s) or filtration systems: sand separator, Fertigation systems (Venturi injector) and chemigation equipment (optional).

- Backwash controller (Backflow prevention device).

- Pressure Control Valve (pressure regulator).

- Distribution lines (main larger diameter pipe, maybe secondary smaller, pipe fittings).

- Hand-operated, electronic, or hydraulic control valves and safety valves.

- Smaller diameter polyethylene tube (often called "laterals").

- Poly fittings and accessories (to make connections).

- Emitting devices at plants (emitter or dripper, micro spray head, inline dripper or inline drip tube).

In drip irrigation systems, pump and valves may be manually or automatically operated by a controller.

Most large drip irrigation systems employ some type of filter to prevent clogging of the small emitter flow path by small waterborne particles. New technologies are now being offered that minimize clogging. Some residential systems are installed without additional filters since potable water is already filtered at the water treatment plant. Virtually all drip irrigation equipment manufacturers recommend that filters be employed and generally will not honor warranties unless this is done. Last line filters just before the final delivery pipe are strongly recommended in addition to any other filtration system due to fine particle settlement and accidental insertion of particles in the intermediate lines.

Drip and subsurface drip irrigation is used almost exclusively when using recycled municipal wastewater. Regulations typically do not permit spraying water through the air that has not been fully treated to potable water standards.

Because of the way the water is applied in a drip system, traditional surface applications of timed-release fertilizer are sometimes ineffective, so drip systems often mix liquid fertilizer with the irrigation water. This is called fertigation; fertigation and chemigation (application of pesticides and other chemicals to periodically clean out the system, such as chlorine or sulfuric acid) use chemical injectors such as diaphragm pumps, piston pumps, or aspirators. The chemicals may be added constantly whenever the system is irrigating or at intervals. Fertilizer savings of up to 95% are being reported from recent university field tests using drip fertigation and slow water delivery as compared to timed-release and irrigation by micro spray heads.

Properly designed, installed, and managed, drip irrigation may help achieve water conservation by reducing evaporation and deep drainage when compared to other types of irrigation such as flood or overhead sprinklers since water can be more precisely applied to the plant roots. In addition, drip can eliminate many diseases that are spread through water contact with the foliage. Finally, in regions where water supplies are severely limited, there may be no actual water savings, but rather simply an increase in production while using the same amount of water as before. In very arid regions or on sandy soils, the preferred method is to apply the irrigation water as slowly as possible.

Pulsed irrigation is sometimes used to decrease the amount of water delivered to the plant at any one time, thus reducing runoff or deep percolation. Pulsed systems are typically expensive and require extensive maintenance. Therefore, the latest efforts by emitter manufacturers are focused on developing new technologies that deliver irrigation water at ultra-low flow rates, i.e. less than 1.0 liter per hour. Slow and even delivery further improves water use efficiency without incurring the expense and complexity of pulsed delivery equipment.

An emitting pipe is a type of drip irrigation tubing with emitters pre-installed at the factory with specific distance and flow per hour as per crop distance.

An emitter restricts water flow passage through it, thus creating head loss required (to the extent of atmospheric pressure) in order to emit water in the form of droplets. This head loss is achieved by friction/turbulence within the emitter.

Advantages and Disadvantages

Drip irrigation and spare drip irrigation tubes in banana farm.

Pot irrigation by On-line drippers.

The advantages of drip irrigation are:

- Fertilizer and nutrient loss is minimized due to a localized application and reduced the leaching.

- Water application efficiency is high if managed correctly.

- Field leveling is not necessary.

- Fields with irregular shapes are easily accommodated.

- Recycled non-potable water can be safely used.

- Moisture within the root zone can be maintained at field capacity.

- Soil type plays a less important role in the frequency of irrigation.

- Soil erosion is lessened.

- Weed growth is lessened.

- Water distribution is highly uniform, controlled by the output of each nozzle.

- Labour cost is less than other irrigation methods.

- Variation in supply can be regulated by regulating the valves and drippers.

- Fertigation can easily be included with minimal waste of fertilizers.

- Foliage remains dry, reducing the risk of disease.

- Usually operated at lower pressure than other types of pressurized irrigation, reducing energy costs.

The disadvantages of drip irrigation are:

- Initial cost can be more than overhead systems.

- The sun can affect the tubes used for drip irrigation, shortening their lifespan.

- The risks of degrading plastic affecting the soil content and food crops. With many types of plastic, when the sun degrades the plastic, causing it to become brittle, the estrogenic chemicals (that is, chemicals replicating female hormones) which would cause the plastic to retain flexibility have been released into the surrounding environment.

- If the water is not properly filtered and the equipment not properly maintained, it can result in clogging or bioclogging.

- For subsurface drip the irrigator cannot see the water that is applied. This may lead to the farmer either applying too much water (low efficiency) or an insufficient amount of water, this is particularly common for those with less experience with drip irrigation.

- Drip irrigation might be unsatisfactory if herbicides or top dressed fertilizers need sprinkler irrigation for activation.

- Drip tape causes extra cleanup costs after harvest. Users need to plan for drip tape winding, disposal, recycling or reuse.

- Waste of water, time and harvest, if not installed properly. These systems require careful study of all the relevant factors like land topography, soil, water, crop and agro-climatic conditions, and suitability of drip irrigation system and its components.

- In lighter soils subsurface drip may be unable to wet the soil surface for germination. Requires careful consideration of the installation depth.

- Most drip systems are designed for high efficiency, meaning little or no leaching fraction. Without sufficient leaching, salts applied with the irrigation water may build up in the root zone, usually at the edge of the wetting pattern. On the other hand, drip irrigation avoids the high capillary potential of traditional surface-applied irrigation, which can draw salt deposits up from deposits below.

- The PVC pipes often suffer from rodent damage, requiring replacement of the entire tube and increasing expenses.

- Drip irrigation systems cannot be used for damage control by night frosts (like in the case of sprinkler irrigation systems).

Drip Tape

Drip tape is a type of thin-walled dripperline used in drip irrigation. The first drip tape was known as "Dew Hose".

Drip tape duct tape is made of polyethylene and is sold flat on reels. The wall thickness typically ranges from 4 to 25 mils (0.1–0.6 mm). Thicker walled tapes are commonly used for permanent subsurface drip irrigation and thinner walled tapes for temporary throw-away type systems in high-value crops.

Water exits from tape through emitters or drippers. The typical emitter spacing ranges from 6 to 24 inches (150–600 mm). In some products, the emitters are manufactured simultaneously with the tape and are actually formed as part of the product itself. In others, the emitters are manufactured separately and installed at the time of production.

Some product is not a tape, but a thin-walled dripperline, but in popular parlance, both types of products are called tapes. Typical tape diameters are 5/8", 7/8", and 1-3/8", with the larger diameters more commonly used on permanent installations with longer runs.

Drip tape is a recyclable material and can be recycled into viable plastic resins for reuse in the plastics manufacturing industry.

Uses

Irrigation dripper.

Drip irrigation is used in farms, commercial greenhouses, and residential gardens. Drip irrigation is adopted extensively in areas of acute water scarcity and especially for crops and trees such as coconuts, containerized landscape trees, grapes, bananas, ber, eggplant, citrus, strawberries, sugarcane, cotton, maize, and tomatoes.

Drip irrigation for garden available in drip kits are increasingly popular for the homeowner and consist of a timer, hose and emitter. Hoses that are 4 mm in diameter are used to irrigate flower pots.

SPRINKLER IRRIGATION

In the sprinkler method of irrigation, water is sprayed into the air and allowed to fall on the ground surface somewhat resembling rainfall. The spray is developed by the flow of water under pressure through small orifices or nozzles. The pressure is usually obtained by pumping. With careful selection of nozzle sizes, operating pressure and sprinkler spacing the amount of irrigation water required to refill the crop root zone can be applied nearly uniform at the rate to suit the infiltration rate of soil.

Advantages of Sprinkler Irrigation

- Elimination of the channels for conveyance, therefore no conveyance loss.

- Suitable to all types of soil except heavy clay.

- Suitable for irrigating crops where the plant population per unit area is very high. It is most suitable for oil seeds and other cereal and vegetable crops.

- Water saving.

- Closer control of water application convenient for giving light and frequent irrigation and higher water application efficiency.

- Increase in yield.

- Mobility of system.

- May also be used for undulating area.

- Saves land as no bunds etc. are required.

- Influences greater conducive micro-climate.

- Areas located at a higher elevation than the source can be irrigated.

- Possibility of using soluble fertilizers and chemicals.

- Less problem of clogging of sprinkler nozzles due to sediment laden water.

Crop Response to Sprinkler

The trials conducted in different parts of the country revealed water saving due to sprinkler system varies from 16 to 70 % over the traditional method with yield increase from 3 to 57 % in different crops and agro climatic conditions.

Response of Different Crops to Sprinkler Irrigation

Crops	Water Saving, %	Yield increase, %
Bajra	56	19

Barley	56	16
Bhindi	28	23
Cabbage	40	3
Cauliflower	35	12
Chillies	33	24
Cotton	36	50
Cowpea	19	3
Fenugreek	29	35
Garlic	28	6
Gram	69	57
Groundnut	20	40
Jowar	55	34
Lucerne	16	27
Maize	41	36
Onion	33	23
Potato	46	4
Sunflower	33	20
Wheat	35	24

General Classification of Different Types of Sprinkler Systems

Sprinkler systems are classified into the following two major types on the basis of the arrangement for spraying irrigation water.

1. Rotating head or revolving sprinkler system.

2. Perforated pipe system.

Rotating head: Small size nozzles are placed on riser pipes fixed at uniform intervals along the length of the lateral pipe and the lateral pipes are usually laid on the ground surface. They may also be mounted on posts above the crop height and rotated through 90°, to irrigate a rectangular strip. In rotating type sprinklers, the most common device to rotate the sprinkler heads is with a small hammer activated by the thrust of water striking against a vane connected to it.

Figure: Example of a few rotating type sprinkler irrigation systems.

Perforated pipe system: This method consists of drilled holes or nozzles along their length through which water is sprayed under pressure. This system is usually designed for relatively low pressure (1 kg/cm²). The application rate ranges from 1.25 to 5 cm per hour for various pressure and spacing.

Based on the portability, sprinkler systems are classified into the following types:

1. Portable system: A portable system has portable main lines, laterals and pumping plant.

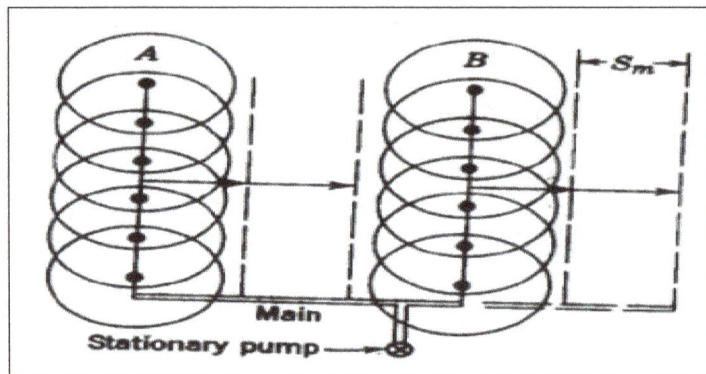

Figure: Fully portable sprinkler irrigation system.

2. Semi portable system: A semi portable system is similar to a portable system except that the location of water source and pumping plant is fixed.

3. Semi permanent system: A semi permanent system has portable lateral lines, permanent main lines and sub mains and a stationery water source and pumping plant.

4. Solid set system: A solid set system has enough laterals to eliminate their movement. The laterals are positions in the field early in the crop season and remain for the season.

5. Permanent system: A fully permanent system consists of permanently laid mains, sub mains and laterals and a stationery water source and pumping plant.

Center Pivot Irrigation

Center-pivot irrigation (sometimes called central pivot irrigation), also called water-wheel and circle irrigation, is a method of crop irrigation in which equipment rotates around a pivot and crops are watered with sprinklers. A circular area centered on the pivot is irrigated, often creating a circular pattern in crops when viewed from above (sometimes referred to as *crop circles*). Most center pivots were initially water-powered, and today most are propelled by electric motors.

Center pivot irrigation is a form of overhead sprinkler irrigation consisting of several segments of pipe (usually galvanized steel or aluminum) with sprinklers positioned along their length, joined together and supported by trusses, and mounted on wheeled towers. The machine moves in a circular pattern and is fed with water from the pivot point at the center of the circle.

For a center pivot to be used, the terrain needs to be reasonably flat; but one major advantage of center pivots over alternative systems that use gravity flow is the ability to function in undulating country. This advantage has resulted in increased irrigated acreage and water use in some areas. The system is in use, for example, in parts of the United States, Australia, New Zealand, and Brazil and also in desert areas such as the Sahara and the Middle East.

Center pivots are typically less than 1600 feet (500 meters) in length (circle radius) with the most common size being the standard 1/4 mile (400 m) machine. A typical 1/4 mile radius crop circle covers about 125 acres of land.

Rotator style pivot applicator sprinkler.   End Gun style pivot applicator sprinkler.

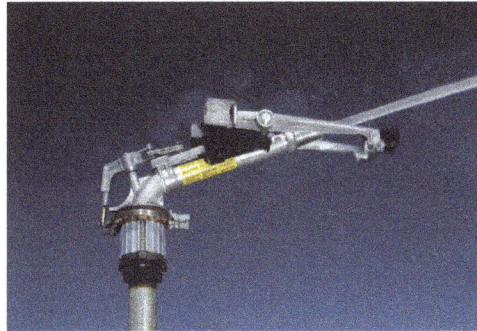

Originally, most center pivots were water-powered. These were replaced by hydraulic systems and electric motor-driven systems. Most systems today are driven by an electric motor mounted at each tower.

The outside set of wheels sets the master pace for the rotation (typically once every three days). The inner sets of wheels are mounted at hubs between two segments and use angle sensors to detect when the bend at the joint exceeds a certain threshold. When the angle is too large, the wheels rotate to keep the segments aligned.

To achieve uniform application, center pivots require an even emitter flow rate across the radius of the machine. Since the outer-most spans (or towers) travel farther in a given time period than the innermost spans, nozzle sizes are smallest at the inner spans and increase with distance from the pivot point. Aerial views show fields of circles created by the watery tracings of "quarter- or half-mile of the center-pivot irrigation pipe," created by center pivot irrigators which use "hundreds and sometimes thousands of gallons a minute."

Center pivot irrigation at Irkhaya Farms in Al Rayyan, Qatar.

Most center pivot systems now have drops hanging from a u-shaped pipe called a *gooseneck* attached at the top of the pipe with sprinkler heads that are positioned a few feet (at most) above the crop, thus limiting evaporative losses and wind drift. There are many different nozzle configurations available including static plate, moving plate and part circle. Pressure regulators are typically installed upstream of each nozzle to ensure each is operating at the correct design pressure.

Drops can also be used with drag hoses or bubblers that deposit the water directly on the ground between crops. This type of system is known as LEPA (Low Energy Precision Application) and is often associated with the construction of small dams along the furrow length (termed furrow diking/dyking). Crops may be planted in straight rows or are sometimes planted in circles to conform to the travel of the irrigation system.

Linear/lateral Move Irrigation Machines

A small center pivot system from beginning to end.

Irrigation equipment can also be configured to move in a straight line, where it is termed a *lateral move*, *linear move*, *wheel move* or *side-roll* irrigation system. In these systems the water is supplied by an irrigation channel running the length of the field. The channel is positioned either at one side or in a line through the center. The motor and pump equipment are mounted on a cart by the supply channel. The cart travels with the machine.

Farmers might choose lateral-move irrigation to keep existing rectangular fields. This can help them convert from furrow irrigation. Lateral-move irrigation is far less common, relies on more complex guidance systems, and requires additional management compared to center pivot irrigation. Lateral-move irrigation is common in Australia. There, systems are usually between 500 and 1,000 meters long.

Benefits

Center-pivot irrigation uses less labor than many other surface irrigation methods, such as furrow irrigation. It also has lower labor costs than ground-irrigation techniques that require digging of channels. Also, center-pivot irrigation can reduce the amount of soil tillage. Therefore, it helps reduce water runoff and soil erosion that can occur with ground irrigation. Less tillage also encourages more organic materials and crop residue to decompose back into the soil. It also reduces soil compaction.

In the United States early settlers of the semiarid High Plains were plagued by crop failures due to cycles of drought, culminating in the disastrous Dust Bowl of the 1930s. Only after World War II when center pivot irrigation became available did the land mass of the High Plains aquifer system transform into one of the most agriculturally productive regions in the world.

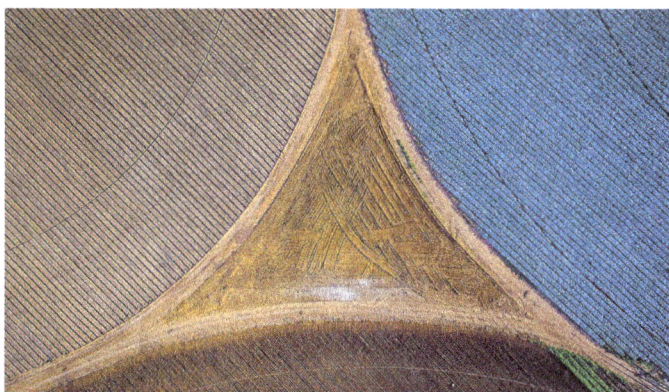

The crops are planted in circles for efficient irrigation.

Negative Effects

Fossil water is a non-renewable resource. Groundwater levels decrease when the rate of extraction by irrigation exceeds the rate of recharge. By 2013 it was shown that as the water consumption efficiency of center-pivot irrigation improved over the years, farmers planted more intensively, irrigated more land, and grew thirstier crops.

In parts of the United States, sixty years of the profitable business of intensive farming using huge center-pivot irrigators has emptied parts of the Ogallala Aquifer (also known as the High Plains Aquifer). One of the world's largest aquifers, it covers an area of approximately 174,000 mi^2 (450,000 km^2) in portions of the eight states of South Dakota, Nebraska, Wyoming, Colorado, Kansas, Oklahoma, New Mexico, and Texas, beneath the Great Plains in the United States.

In 1950, irrigated cropland covered 250,000 acres. With the use of center-pivot irrigation, nearly three million acres of land were irrigated in Kansas alone. At some places, during maximum extraction, the water table dropped more than five feet (1.5 m) per year. In extreme cases, wells had to be greatly deepened to reach the steadily falling water table. In some places in the Texas Panhandle, the water table has been drained (dewatered). "Vast stretches of Texas farmland lying over the aquifer no longer support irrigation. In west-central Kansas, up to a fifth of the irrigated farmland along a 100-mile (160 km) swath of the aquifer has already gone dry." It would take hundreds to thousands of years of rainfall to replace the groundwater in the dried up aquifer.

DEFICIT IRRIGATION

Deficit irrigation (DI) is a watering strategy that can be applied by different types of irrigation application methods. The correct application of DI requires thorough understanding of the yield response to water (crop sensitivity to drought stress) and of the economic impact of reductions in harvest. In regions where water resources are restrictive it can be more profitable for a farmer to maximize crop water productivity instead of maximizing the harvest per unit land. The saved water can be used for other purposes or to irrigate extra units of land. DI is sometimes referred to as incomplete supplemental irrigation or regulated DI.

Crop Water Productivity

Crop water productivity (WP) or water use efficiency (WUE) expressed in kg/m³ is an efficiency term, expressing the amount of marketable product (e.g. kilograms of grain) in relation to the amount of input needed to produce that output (cubic meters of water). The water used for crop production is referred to as crop evapotranspiration. This is a combination of water lost by evaporation from the soil surface and transpiration by the plant, occurring simultaneously. Except by modeling, distinguishing between the two processes is difficult. Representative values of WUE for cereals at field level, expressed with evapotranspiration in the denominator, can vary between 0.10 and 4 kg/m³.

Experiences with Deficit Irrigation

For certain crops, experiments confirm that DI can increase water use efficiency without severe yield reductions. For example for winter wheat in Turkey, planned DI increased yields by 65% as compared to winter wheat under rainfed cultivation, and had double the water use efficiency as compared to rainfed and fully irrigated winter wheat. Similar positive results have been described for cotton. Experiments in Turkey and India indicated that the irrigation water use for cotton could be reduced to up to 60 percent of the total crop water requirement with limited yield losses. In this way, high water productivity and a better nutrient-water balance was obtained.

Certain Underutilized and horticultural crops also respond favorably to DI, such as tested at experimental and farmer level for the crop quinoa. Yields could be stabilized at around 1.6 tons per hectare by supplementing irrigation water if rainwater was lacking during the plant establishment and reproductive stages. Applying irrigation water throughout the whole season (full irrigation) reduced the water productivity. Also in viticulture and fruit tree cultivation, DI is practiced.

Scientists affiliated with the Agricultural Research Service (ARS) of the USDA found that conserving water by forcing drought (or deficit irrigation) on peanut plants early in the growing season has shown to cause early maturation of the plant yet still maintain sufficient yield of the crop. Inducing drought through deficit irrigation earlier in the season caused the peanut plants to physiologically "learn" how to adapt to a stressful drought environment, making the plants better able to cope with drought that commonly occurs later in the growing season. Deficit irrigation is beneficial for the farmers because it reduces the cost of water and prevents a loss of crop yield (for certain crops) later on in the growing season due to drought. In addition to these findings, ARS scientists suggest that deficit irrigation accompanied with conservation tillage would greatly reduce the peanut crop water requirement.

For other crops, the application of deficit irrigation will result in a lower water use efficiency and yield. This is the case when crops are sensitive to drought stress throughout the complete season, such as maize.

Apart from university research groups and farmers associations, international organizations such as FAO, ICARDA, IWMI and the CGIAR Challenge Program on Water and Food are studying DI.

Reasons for Increased Water Productivity under Deficit Irrigation

If crops have certain phenological phases in which they are tolerant to water stress, DI can increase the ratio of yield over crop water consumption (evapotranspiration) by either reducing the water loss by unproductive evaporation, and/or by increasing the proportion of marketable yield to the totally produced biomass (harvest index), and/or by increasing the proportion of total biomass production to transpiration due to hardening of the crop - although this effect is very limited due to the conservative relation between biomass production and crop transpiration, - and/or due to adequate fertilizer application and/or by avoiding bad agronomic conditions during crop growth, such as water logging in the root zone, pests and diseases, etc.

Advantages

The correct application of DI for a certain crop:

- Maximizes the productivity of water, generally with adequate harvest quality.

- Allows economic planning and stable income due to a stabilization of the harvest in comparison with rainfed cultivation.

- Decreases the risk of certain diseases linked to high humidity (e.g. fungi) in comparison with full irrigation.

- Reduces nutrient loss by leaching of the root zone, which results in better groundwater quality and lower fertilizer needs as for cultivation under full irrigation.

- Improves control over the sowing date and length of the growing period independent from the onset of the rainy season and therefore improves agricultural planning.

Constraints

A number of constraints apply to deficit irrigation:

- Exact knowledge of the crop response to water stress is imperative.

- There should be sufficient flexibility in access to water during periods of high demand (drought sensitive stages of a crop).

- A minimum quantity of water should be guaranteed for the crop, below which DI has no significant beneficial effect.

- An individual farmer should consider the benefit for the total water users community (extra land can be irrigated with the saved water), when he faces a below-maximum yield.

- Because irrigation is applied more efficiently, the risk for soil salinization is higher under DI as compared to full irrigation.

Modeling

Field experimentation is necessary for correct application of DI for a particular crop in a particular region. In addition, simulation of the soil water balance and related crop growth (crop water productivity modeling) can be a valuable decision support tool. By conjunctively simulating the effects of different influencing factors (climate, soil, management, crop characteristics) on crop production, models allow to better understand the mechanism behind improved water use efficiency, to schedule the necessary irrigation applications during the drought sensitive crop growth stages, considering the possible variability in climate, to test DI strategies of specific crops in new regions, and to investigate the effects of future climate scenarios or scenarios of altered management practices on crop production.

ADVANTAGES AND DISADVANTAGES OF IRRIGATION

Advantages of Irrigation:

1. For proper nourishment of crops certain amount of water is required. If rainfall is insufficient there will be deficiency in fulfillment of water requirement. Irrigation tries to remove this deficiency caused due to inadequate rainfall. Thus, irrigation comes to rescue in dry years.

2. Irrigation improves the yield of crops and makes people prosperous. The living standards of the people is thereby improved.

3. Irrigation also adds to the wealth of the country in two ways. Firstly as bumper crops are produced due to irrigation it makes country self-sufficient in food requirements. Secondly as the irrigation water is taxed when it is supplied to the cultivators, it adds to the revenue.

4. Irrigation makes it possible to grow cash crops which give good returns to the cultivators than the ordinary crops they might have grown in absence of irrigation. Fruit gardens, sugarcane, potato, tobacco etc., are the cash crops.

5. Sometimes large irrigation channels can be used as a means of communication.

6. The falls which come across the irrigation channels can be utilised for producing hydroelectric power.

7. Domestic advantages should not be overlooked. Irrigation facilitates bathing, cattle watering etc., and improves freshwater circulation.

8. Irrigation improves the groundwater storage as water lost due to seepage adds to the groundwater storage.

9. Along the banks of large irrigation channels plantation can be successfully done which not only helps introducing social forestry but also improves environmental status of the region.

10. New irrigation works are started at the time of famines to provide employment to a large number of population. These works are called famine works or relief works.

11. When watering facility is provided to a barren land, the value of this land gets appreciated.

Disadvantages of Irrigation:

1. Excessive seepage and leakage of water forms marshes and ponds all along the channels. The marshes and the ponds in course of time become the colonies of the mosquito, which gives rise to a disease like malaria.

2. Excessive seepage into the ground raises the water-table and this in turn completely saturates the crop root-zone. It causes waterlogging of that area.

3. It lowers the temperature and makes the locality damp due to the presence of irrigation water.

4. Under irrigation canal system valuable residential and industrial land is lost.

5. Initial cost of irrigation project is very high and thereby the cultivators have to pay more taxes in the form of levy.

6. Irrigation works become obstacles in the way of free drainage of water during rainy season and thus results in submerging standing crops and even villages.

References

* Morris, John Miller (2003). Sherry L. Smith (ed.). The Future of the Southern Plains. Norman, Oklahoma: University of Oklahoma Press. P. 275. ISBN 0806137355

* Irrigation, entry: newworldencyclopedia.org, Retrieved, 17 February, 2019

* El-Dine, T. G.; Hosny, M. M. (2000). "Field evaluation of surge and continuous flows in furrow irrigation systems". Water Resources Management. 14 (2): 77–87. Doi:10.1023/a:1008189004992

* Horst, M. G.; Shamutalov, S. S.; Goncalves, J. M.; Pereira, L. S. (2007). "Assessing impacts of surge-flow irrigation on water saving and productivity of cotton". Agricultural Water Management. 87 (2): 115–127. Doi:10.1016/j.agwat.2006.06.014

* "DEW-HOSE Trademark - Registration Number 0847046 - Serial Number 72249303 :: Justia Trademarks". Trademarks.justia.com. Retrieved 2016-06-12

* Bainbridge, David A (June 2001). "Buried clay pot irrigation: a little known but very efficient traditional method of irrigation". Agricultural Water Management. 48 (2): 79–88. Doi:10.1016/S0378-3774(00)00119-0

* India, Press Trust of (2006-05-03). "Jain Irrigation buys Chapin for $6 mn". Business Standard India. Retrieved 2017-09-30

* Spring_irrigation, agricultural_engineering: tnau.ac.in, Retrieved, 17 February, 2019

* Morgan, Robert (1993). Water and the Land. Cathedral City, CA: Adams Publishing Corp. Pp. 35–36. ISBN 0935030026

* Major-advantages-and-disadvantages-of-irrigation, irrigation: yourarticlelibrary.com, Retrieved 28 February, 2019

Permissions

We would like to thank the editorial team for lending their expertise to make the book truly unique. They have played a crucial role in the development of this book. Without their invaluable contributions this book wouldn't have been possible. They have made vital efforts to compile up to date information on the varied aspects of this subject to make this book a valuable addition to the collection of many professionals and students.

This book was conceptualized with the vision of imparting up-to-date and integrated information in this field. To ensure the same, a matchless editorial board was set up. Every individual on the board went through rigorous rounds of assessment to prove their worth. After which they invested a large part of their time researching and compiling the most relevant data for our readers.

The editorial board has been involved in producing this book since its inception. They have spent rigorous hours researching and exploring the diverse topics which have resulted in the successful publishing of this book. They have passed on their knowledge of decades through this book. To expedite this challenging task, the publisher supported the team at every step. A small team of assistant editors was also appointed to further simplify the editing procedure and attain best results for the readers.

Apart from the editorial board, the designing team has also invested a significant amount of their time in understanding the subject and creating the most relevant covers. They scrutinized every image to scout for the most suitable representation of the subject and create an appropriate cover for the book.

The publishing team has been an ardent support to the editorial, designing and production team. Their endless efforts to recruit the best for this project, has resulted in the accomplishment of this book. They are a veteran in the field of academics and their pool of knowledge is as vast as their experience in printing. Their expertise and guidance has proved useful at every step. Their uncompromising quality standards have made this book an exceptional effort. Their encouragement from time to time has been an inspiration for everyone.

The publisher and the editorial board hope that this book will prove to be a valuable piece of knowledge for students, practitioners and scholars across the globe.

Index